THE STUDY OF

INSTINCT

FIG. 1. Male Three-spined Stickleback (*Gasterosteus aculeatus*) in threat posture in front of mirror. This activity is innate (p. 51), dependent on internal (motivational, p. 63) and external (sensory, p. 28) factors. It has an intimidating effect on other males of the same species (p. 177). Historically, it is displacement sand-digging (p. 113), changed by ritualization (p. 184)

THE STUDY OF INSTINCT

BY

N. TINBERGEN

PROFESSOR OF ANIMAL BEHAVIOUR
IN THE UNIVERSITY OF OXFORD

WITH A NEW INTRODUCTION

OXFORD
AT THE CLARENDON PRESS

Oxford University Press, Ely House, London W.1

GLASGOW NEW YORK TORONTO MELBOURNE WELLINGTON
CAPE TOWN SALISBURY IBADAN NAIROBI LUSAKA ADDIS ABABA
BOMBAY CALCUTTA MADRAS KARACHI LAHORE DACCA
KUALA LUMPUR SINGAPORE HONG KONG TOKYO

FIRST EDITION 1951
SECOND IMPRESSION 1952

REPRINTED LITHOGRAPHICALLY IN GREAT BRITAIN
BY BUTLER & TANNER LTD, FROME AND LONDON
FROM SHEETS OF THE SECOND IMPRESSION
1955, 1958, 1969 (WITH A NEW INTRODUCTION)

INTRODUCTION TO 1969 REPRINT

A FEW words of justification are required for the decision to reprint unaltered a text written twenty years ago. In my original preface I said, somewhat apologetically: 'Someone had to make the attempt' [at a coherent presentation of ethology], and also: 'I cannot imagine a better result of my venture than that it should lead to better presentations in the near future.' I could not have foreseen that the behavioural sciences would so soon show a truly explosive growth, nor that, in spite of this growth (which *has* led to much better presentations), there would still be a demand for my original book two decades after it was published.

This demand has put me in a difficult dilemma. On the one hand, copyright laws being what they are, I am told that I cannot prevent some keen publisher from reprinting the old text. On the other, I feel that the original book, having rendered its service, ought to be relegated to the shelves for books of merely historical interest. Having neither the time nor the inclination to do the sensible thing and write an entirely new text, I have chosen, rightly or wrongly, to allow reprinting of the original text, preceded merely by a few words of caution. This compromise solution seems to me acceptable because, while much in the book is now outdated, I feel that its overall approach to behavioural problems is still representative of what I like to call the biological study of behaviour. The aim of this new preface is therefore to help the reader in assessing the old text critically, yet with the sympathetic attention I feel it still deserves.

I still consider the main merits of the book its brevity, and the avoidance of needless technical jargon. Moreover, in its scope it occupies a useful middle position between very brief introductions such as on the one hand those by Dethier and Stellar (1961), Manning (1967), Klopfer and Hailman (1967), and Tinbergen (1965*b*); and on the other, solid texts such as those by Thorpe (1963), Hinde (1966), Marler and Hamilton (1967), and Eibl-Eibesfeldt (1967).

A comparison of these books, amongst themselves and with mine, shows how much progress has been made. But the fact that they all differ so strikingly in their conception, their structuring, and even their contents also reveals the immature, yet refreshingly exploratory nature of the stage now reached.

I think that it will be useful if I indicate, as briefly as I can, in which directions I feel that progress has been most striking, giving no more than a very few representative references.

First of all, no reader can fail to notice how superficial my descriptions of behaviour were at the time; how little information was given about the actual movements which formed the inductive basis of my treatise, and in particular how little was said about the temporal patterning of the diversity of motor patterns shown by any animal. The accuracy of this descriptive work is now rapidly improving. This is in part due to the increased use of recording techniques, including the cine-camera and a variety of event recorders. A special development is noticeable in the recording of sounds and their representation as sound spectograms, which render a number of parameters visible, and also reproducible in print. In addition, the awareness (emphasized in the original text) of the need of adapting one's descriptions to one's aims, to the questions one has in mind, is becoming increasingly evident in the descriptive parts of many publications—descriptions as such, 'ethograms' without open acknowledgement of analytical aims, are now less common than they were. Yet this more 'problem-oriented' approach has in my view its dangers. We can apply to ethology what F. A. Beach once said of American psychology: that 'in its haste to step into the twentieth century' it had tried to rush through the preparatory inductive, descriptive phase—a thing no natural science can afford to do. Having myself always spent long periods of exploratory watching of natural events, of pondering about what exactly it was in the observed behaviour that I wanted to understand before developing an experimental attack, I find this tendency of prematurely plunging into quantification and experimentation, which I observe in many younger workers, really disturbing, unless, as happens to some, they do, from time to time, return, more purposefully than before, to plain, though more sophisticated, watching. A man who follows this course most beautifully is K. D. Roeder, whose gradually unfolding work on hearing in moths and the interaction between moths and bats is to me a model of this different way of balancing observation and experiment.

In spite of these misgivings I must, of course, acknowledge that striking developments have also occurred in what I should like to call the pre-experimental analysis of observational data. Progress is most noticeable in two main areas. Where a certain degree of order has to be detected in the seemingly chaotic sequences of behaviour patterns of intact animals, various methods of detecting correlations in the appearance of recognizable elements of behaviour have been applied. The main merits of, for instance, factor analysis as applied by Wiepkema (1961) and cross-correlation methods such as applied by Delius (1968) seem to me to lie in the fact that they provide hard rather than intuitive evidence for the assumption that, within the total behaviour repertoire of an animal, certain groupings can be distinguished; groupings which indicate that these elements are in some way causally connected. The nature of these causal connections will, of course, as always, have to be ultimately unravelled by experiments.

A second development of great promise is the application of cybernetics in the analysis of behavioural mechanisms. Its usefulness lies in detecting where the unexplored cause-effect links are to be found that have to be analysed in order to understand how a behavioural mechanism works. This value is being beautifully demonstrated by Hassenstein (1966).

These methods are undoubtedly giving more purpose and direction to the real causal analysis of behavioural mechanisms. In this investigation of 'what makes behaviour happen the way it does' (as distinct from the equally interesting question how the effects of behaviour influence the success of the animal), I remain convinced that it is useful, at least in the present state of our knowledge, to 'dissect' the master question into three stages. Stage one investigates the causation of repetitive, short-term cycles within the life span of the individual (for instance the causation of feeding acts, or feeding bouts, or, on an even longer time scale, the recurrence of reproductive cycles). The second stage is concerned with the development of behaviour in the individual, which forms one single, long cycle: the individual's life span. The third stage is the study of evolution, the process resulting from the series of changes in the ontogenies of many successive generations.

In the first task, that of unravelling the 'moment-to-moment' control of recurring behavioural cycles, the natural course, foreshadowed by Chapters II–V inclusive, is that of bridging the gap between the analysis of observed behaviour as carried out by ethologists and that of component mechanisms, which is conventionally considered the task of physiologists. Two general points can be made here. Firstly, it has become clear that the gap between the two fields of study is very much wider than, perhaps naïvely, we imagined twenty years ago. Secondly, however, bridges across this gap are being built in a great variety of ways, and with an almost bewildering diversity of techniques. Ethologists and psychologists—in this field the distinction fades—are penetrating into the 'machinery' underlying complex behaviour systems. As examples I should like to mention the studies of Hinde and of Lehrman on the reproductive behaviour of birds, the gist of which is summarized in Hinde's book.

Physiologists, conventionally analysing relatively minor components of such systems, are gradually turning their attention to more highly integrated phenomena. Resisting, as I have to, the temptation to go into details, I could quote as one example (which happens to concern the integration of visual input in a decapod eye) the work by Wiersma, Waterman, and Bush (1961).

Working on various levels between these two extremes of approach are a number of scientists whom it would be impossible to label as either ethologists or physiologists; as one out of numerous examples I select the paper by von Holst and v. Saint Paul (1963), which I consider an important milestone in the rapidly growing field of brain stem stimulation. This method, particularly when supplemented by conventional recording of action potentials and by pharmacological techniques, may well grow into an important part of the bridge between ethology and neurophysiology, provided it does not lose its contact (as it is in danger of doing) with the full behaviour of the intact animal.

In view of these developments, on which I cannot possibly elaborate here, the reader should be particularly alert when reassessing Chapter V. Much of what I wrote there obviously contains a core of truth; thus the basic idea of distinguishing groups of behaviours that are both causally and functionally linked (the 'instincts' in more or less the same sense as von Uexkull's 'Funktionskreise') seems to be undoubtedly sound; and so is the idea of a hierarchical organization. In my view the main shortcoming of my original presentation is that it ignores the self-regulatory nature of behaviour, and consequently does not even raise the question of feedbacks. Modern research begins to show the heuristic value of a systematic search for feedbacks, both negative and positive, wherever the achievements of the as yet largely unknown mechanisms tell us that they *must* be operative. As a secondary result of this widened approach the old notions about the causation of displacement activities and other results of 'motivational conflicts' are being re-examined (see, for instance, Hinde's treatment).

Studies of behaviour ontogeny have also entered a new phase. The most important advances seem to me the almost universal acceptance of the fact that many behaviour patterns can develop, some to a high degree of complexity, prior to interaction with the environment of the types collectively called 'learning'; further the acknowledgement that, in spite of this, a rigid dichotomy between 'innate behaviour' and 'learnt behaviour' is no more than a first hesitant step in the analysis of the developmental process as a whole; and, thirdly, that even where learning plays an important part in the development of behaviour, there is often, if not always, a selectiveness in what is learned and what not—a bias which guides the interaction with the environment—and that this bias can itself have been internally programmed. Here Lorenz's concept of the 'Innate Teaching Mechanism' (1965), a concept akin to what Thorpe calls 'Templates', seems to me to have great heuristic value, particularly if, as I have proposed (1968), Lorenz's concept is widened to one of 'Internal Teaching Mechanisms'.

Studies of the evolution of behaviour naturally fall into three parts: (1) the descriptive reconstruction of the course of evolution that has led to the present situation; (2) the genetics proper of behaviour; and (3) the study of directional changes under the influence of natural selection. These fields too are rapidly developing. When I look back at Chapters VII and VIII, I feel that these parts are at the same time the most valuable of the original text, and the least mature. More than the other parts they contribute to that 'breath of fresh air' that critics have referred to when ethology first appeared on the British scene; more than the other parts they show the potential of studying animals in their natural environment, i.e. in the environment that exerts the pressures which each animal species has to meet.

The immaturity of my original treatment, and even of modern studies in this field, are undoubtedly due to the fact that the development of the behavioural, and indeed the biological sciences in general, is taking a one-sided course in that a disproportionally great effort is channelled into questions of causation of behaviour. I feel very strongly that an equally intense effort ought to be made to understand the *effects* of behaviour; of the ways in which it influences the survival of the species; and that we should try much harder to understand the state of adaptedness and the process of evolutionary adaptation, if for no other reason than that this has led to the structuring of our own behaviour. Adaptation has been steered by the environment with its innumerable selection pressures; one has to agree with Mayr (1963) that 'the environment is one of the most important evolutionary factors' (p. 7); but also with his statement that (p. 9) 'There are vast areas of modern biology, for instance . . . the study of behavior, in which the application of evolutionary principles is still in the most elementary stage.' It is, among other reasons, because I agree with these statements that I still adhere to what I wrote in 1965: 'The naturalist knows perhaps better than any other zoologist how immensely complex are the relationships between an animal and its environment, how numerous and how severe are the pressures the environment exerts, the challenges the animal has to meet in order not merely to survive, but also to contribute substantially to future generations. *He also realizes how little we really know*' (p. 522). (Italics mine, N.T.)

This gap in our science is the more remarkable, since it has been clearly shown that the contribution to survival of behaviour can be studied experimentally just as well as that of structures, of which the classical example, the external colouration of animals, has already been subjected to such experimental study for a long time.

It is tempting to ponder this over-emphasis on studies of causation. I believe that it is partly due to the fact that, as the development of physics and chemistry have shown, knowledge of the causes underlying natural events provides us with the power to manipulate these events and 'bully them into subservience'. It is perhaps for this reason that Man, and particularly urban Man, is inclined even in his biological studies to ape physics, and so to contribute to the satisfaction of his urge to conquer nature. In the process he seems to forget that Science has to aim primarily at understanding, and not merely at power, which, however vital to us, is merely one of the results of scientific understanding. Paradoxically, however, the unprecedented degree of power over natural events which we have achieved carries in its wake a dangerous consequence; we have broken out of our ecological niche into the niches of almost all other species, and have thus changed our environment (including our social environment) out of all recognition. As a result our behavioural organization is no longer faced with the environment in which this organization was moulded and, as a consequence, misfires. These disruptive consequences of our behaviour now threaten the very existence of our species: pollution and depletion of our natural resources, our population explosion, our stressful social environment, the threat of nuclear war are all consequences of misfiring of our behaviour. An increasing, but still far too small number of people begin to realize that we are caught in a vicious circle: the very success of our behaviour has led to a situation from which only a better understanding and controlled change of our behaviour can extract us.

It seems to me important to realize that the catchphrase 'misfiring of our behaviour', when more sharply defined, refers to the *relation between our behaviour and our environment*, and that it is therefore this relation, not just behaviour in isolation, that we have to study. Most animals still do live in their natural, or near-natural environments, which is the one thing we definitely do not do. Evolutionary studies show us the way towards fuller understanding of the relations between environment and behaviour, and it is therefore to evolutionary studies, including of Man himself, that we shall have to turn. It is ultimately because in almost all modern texts the ecological and evolutionary aspects of behaviour are under-represented that I decided that reprinting of my original text was still worth while. In spite of its *naïveté*, perhaps through this very *naïveté*, it may well help readers even now to see the modern developments of our science in a wider biological context.

N. TINBERGEN

Oxford
February 1969

REFERENCES MENTIONED IN THE INTRODUCTION

DELIUS, J. D., 1969: A stochastic analysis of the maintenance behaviour of skylarks. *Behav.* **33**, 137–78.

DETHIER, V. G., and E. STELLAR, 1961: *Animal Behavior.* Englewood Cliffs, N.J.

EIBL-EIBESFELDT, I., 1967: *Grundriss der vergleichenden Verhaltensforschung.* Muenchen.

HASSENSTEIN, B., 1966: *Kybernetik und Biologische Forschung.* Frankfurt/Main.

HINDE, R. A., 1966: *Animal Behaviour.* New York–London (2nd edition in the press).

VON HOLST, E., and U. VON SAINT PAUL, 1963: On the functional organisation of drives. *Anim. Behav.* **11**, 1–20.

KLOPFER, P. H., and J. P. HAILMAN, 1967: *An Introduction to Animal Behavior.* Englewood Cliffs, N.J.

LORENZ, K., 1965: *Evolution and Modification of Behavior.* Chicago–London.

MARLER, P. R., and W. J. HAMILTON, 1967: *Mechanisms of Animal Behavior,* New York.

MANNING, A., 1967: *An Introduction to Animal Behaviour.* London.

MAYR, E., 1963: *Animal Species and Evolution.* London.

ROEDER, K. D., 1964: Aspects of the noctuid tympanic nerve response having significance in the avoidance of bats. *J. Ins. Physiol.* **10**, 529–46.

THORPE, W. H., 1963: *Learning and Instinct in Animals.* London.

TINBERGEN, N., 1965*a*: Behavior and Natural Selection, in J. A. Moore (ed.) *Ideas in Modern Biology.* New York, 521–42.

—— 1965*b*: *Animal Behavior.* Life Nature Library, New York.

—— 1968: On war and peace in animals and Man. *Science,* **160,** 1411–18.

WIEPKEMA, P. R., 1961: An ethological analysis of the reproductive behaviour of the Bitterling. *Arch. néerl. Zool.* **14,** 103–99.

WIERSMA, C. A. G., T. H. WATERMAN, and B. M. H. BUSH, 1961: Impulse traffic in the optic nerve of decapod Crustacea. *Science,* **134,** 1435.

PREFACE

THIS book is an extension of a series of lectures delivered at New York in February 1947 under the auspices of the American Museum of Natural History and Columbia University.

It is a programme rather than an exhaustive treatise of results. Few will realize its shortcomings more fully than I do. Nevertheless, publication does not seem to me to be premature. The study of instinct was given fresh impetus in the 1930's when K. Lorenz, building upon foundations laid by Heinroth and others, opened new lines of approach to the ever-challenging problem of innate behaviour. Since that time many workers, especially on the European continent, have followed Lorenz's lead and laid the foundation of animal ethology. The results obtained thus far, though still highly fragmentary, have proved the new, and yet old, objective method of tackling instinct to be highly fertile.

My presentation has a dual aim. First, it is intended to call the attention of Anglo-American workers to research done on the European continent. Almost all of this work has been published in the German language, and much of it has not penetrated into English and American science. By presenting a review of this work in the English language I hope to contribute to international co-operation in the science of animal behaviour. Second, this book is an attempt at an organization of the ethological problems into a coherent whole. This applies especially to the problems of the causes underlying instinctive behaviour. These problems are dealt with in Chapters I to V inclusive, in which a systematic treatment is attempted. My principal aims in this part have been: (1) to elucidate the hierarchical nature of the system of causal relations, and to stress the paramount importance of recognizing the different levels of integration; and (2) to bring ethology into contact with neurophysiology.

Chapters VI, VII, and VIII, dealing respectively with ontogeny, adaptiveness, and evolution, should not be considered attempts at such a systematic treatment. I added them after considerable hesitation. The fragmentary and more or less unbalanced nature of these chapters, while due in part to my own shortcomings, are also due to the unsatisfactory state of these more or less neglected fields of our science. My main motive for including these chapters has been the hope that by doing so I could contribute towards a more harmonious development of ethology as a whole.

I am fully aware that this first attempt at a synthetic treatment is incomplete and unsatisfactory in many respects. But somebody had to make the attempt. I cannot imagine a better result of my venture than that it should lead to better presentations in the near future.

I wish to thank J. J. A. van Iersel, Prof. C. J. van der Klaauw, Dr. David Lack, Dr. Ernst Mayr, L. de Ruiter, W. M. S. Russell, and Dr. J. Verwey for reading and criticizing parts of the manuscript, D. Lehrman and Prof. P. B. Medawar for revising the English text, and P. H. Creutzberg and P. Sevenster for preparing most of the illustrations.

The manuscript was completed by Christmas 1948. The decision to transfer printing and publishing from the U.S.A. to England has delayed the appearance of the work in book form.

My thanks are further due to the following authors and publishers for the permission to use figures published in their works: Prof. G. P. Baerends, Prof. F. A. Beach, Dr. G. van Beusekom, Dr. Fr. Brock, Dr. M. Brügger, Prof. W. von Buddenbrock, Dr. G. E. Coghill, Prof. Sv. Dijkgraaf, Mr. and Mrs. Duym-van Oyen, Prof. K. von Frisch, Prof. E. von Holst, Dr. V. Holstein, Mr. F. L. Jaques, Prof. H. S. Jennings, Prof. O. Koehler, Dr. A. Kortlandt, Prof. A. Kühn, Dr. D. Lack, Dr. H. Laven, Dr. H. W. Lissmann, Prof. K. Lorenz, Mr. G. F. Makkink, Dr. O. Mast, Mrs. M. M. Nice, Prof. A. S. Pearse, Dr. A. Seitz, Lt.-Col. W. P. R. Tenison, Dr. L. Tinbergen, Prof. P. Weiss; and the American Museum of Natural History for *Natural History*, the American Philosophical Society for the *Proceedings of the American Philosophical Society*, Messrs. Ernest Benn for Norman, *A History of Fishes*; the Biologische Anstalt Helgoland for *Wissensch. Meeresunters. Abt. Helgoland*; Messrs. E. J. Brill for *Behaviour* and for *Ardea*; the Cambridge University Press for Coghill, *Anatomy and the Problem of Behaviour*, and for the *Journal of Experimental Biology*; the Carnegie Institution of Washington for the C.I.W. publications; the Columbia University Press for Jennings, *The Behavior of the Lower Organisms*; the Deutsche Ornithologische Gesellschaft for the *Journal für Ornithologie*; Messrs. Gustav Fischer for Meisenheimer, *Geschlecht und Geschlechter im Tierreich*, for Kühn, *Die Orientierung der Tiere im Raum*, and for Brun, *Die Raumorientierung der Ameisen*; G. E. C. Gads Forlag for Holstein, *Fiskehejren*; Messrs. Paul Hoeber for *Psychosomatic Medicine*; the Linnean Society of New York for Nice, *Studies in the Life History of the Song Sparrow*, i; the Musée d'Histoire Naturelle in Genève for *Revue suisse de Zoologie*; the Nederlandse Entomologische Vereniging for *Tijdschrift voor Entomologie*; Messrs. Paul Parey for *Zs. f. Tierpsychologie*; Messrs. Benno Schwabe A. G. for *Helv. Physiol. et Pharmacol. Acta*; Schweizerbartsche Verlagsbuchhandlung for *Zoologica Wien*; the Smithsonian Institution of Washington for the *Annual Reports Smithsonian Institution, Washington*; Messrs. Julius Springer for *Zs. für vergleichende Physiol.*, for *Pflügers Archiv* and for Von Frisch, *Aus dem*

Leben der Bienen; Stazione Zoologica Naples for *Pubbl. della Staz. Zool. di Napoli*; Georg Thieme Verlag for the *Biol. Zentralblatt* and for Kühn, *Zoologie im Grundriss*; Stichting Trekstation Texel for the *Jaarverslagen*; Messrs. J. Versluys for *De Levende Natuur*; Messrs. Wiley & Sons, Inc., for Mast, *Light and the Behavior of Organisms*; Messrs. Williams & Wilkins for the *Comparative Psychol. Monographs*; and Messrs. Witherby for Lack, *The Life of the Robin*.

N. T.

OXFORD
November 1950

CONTENTS

ILLUSTRATIONS

I

ETHOLOGY: THE OBJECTIVE STUDY OF BEHAVIOUR

INTRODUCTION

THE frontispiece shows a sexually active male three-spined stickleback (*Gasterosteus aculeatus*) reacting to (that is, fighting) its reflection in a mirror. This response is shown only towards other males of the same species who trespass on his territory.

This observation, like all other observations of animal behaviour, however trivial they may seem to be, gives rise to the question upon which the scientific study of behaviour, or ethology, is based: Why does the animal behave as it does?

This question covers a rather complex set of problems. One of them is to determine the causal structure underlying the behaviour. As we shall see, in this particular case one factor is a special kind of visual stimulus. The simple test demonstrated by the frontispiece, namely, presenting the animal with a mirror, shows the visual nature of the stimulus and represents the first step in its study.

Another causal factor is of an internal nature. The fact that this response is shown only by males in reproductive condition suggests that sex hormones may have something to do with it. If this should be proved by experiments, it would represent another step towards the solution of the causal problems.

Both external stimuli and hormones ordinarily exert their influence on the behaviour through the nervous system. Therefore, our next task after the study of the sensory stimulus and the hormone involved would have to be a study of what happens in the nervous system. When we make an accurate description of the particular type of behaviour shown in the present case, we shall see that a great many muscles are involved, each of which contracts (1) to a certain degree, and (2) in co-ordination with the other muscles. This indicates the complexity of the nervous processes involved in this single action.

Thus the problem of the causal structure underlying the behaviour leads to a study of the functions of the sense organs, of hormones, of the nervous system, of the muscles, and, particularly, of the co-ordination of these functions, of their integration into the act of behaviour as a whole.

However, if we should solve this particular problem—that is to say, if we discovered the physiological mechanism at work in the behaving animal—we should still have answered only part of the whole causal question. For, since the full-grown animal is the outcome of a developmental process beginning with the unicellular egg, the description and causal explanation of its development are natural extensions of our causal study. In other words, it is part of our task to study the ontogeny of behaviour as well.

Even this extension of ethology does not cover the whole field of the causal study of behaviour. We also want to know how the animals that possess these mechanisms of growth and of adult functioning have evolved in the course of history, as we know they must have done. Thus, willingly or unwillingly, we are led on to the study of the evolution of behaviour. In spite of the special difficulties that meet us here, difficulties inherent in attempting to trace any historical process, we must consider the study of evolution as a natural outgrowth of the causal study of present behaviour.

To return to our basic question, 'Why?', it is obvious that it covers more than the problem of causes. When we see by experiments, as we shall see later, that the peculiar behaviour of the stickleback male results in frightening other males away, we conclude that it behaves in this way in order to defend its territory against other males. Or, to put it more objectively, the biological significance of its behaviour is that it drives off other males. Pointing in this way to the function an activity serves is also a way of answering the question, 'Why?'

Thus the basic problem splits itself into several more or less separate problems, each the subject-matter of special fields of science. The study of the mechanisms at work in an animal in action is connected with sensory physiology and especially neurophysiology, with endocrinology, and, to a lesser extent, with muscle physiology. It culminates in the study of co-ordination, of integration. The study of evolution is connected with taxonomy, with ecology, and with genetics. The study of the functions of behaviour has ecological and sociological aspects.

In this work I shall concentrate on only part of this complex of problems, and while of course it would be undesirable entirely to neglect any of the fields mentioned, I shall deal mainly with the causes of innate behaviour.

AIMS, SCOPE, AND LIMITATIONS OF A CAUSAL STUDY OF INNATE BEHAVIOUR

Because our purpose is the study of the causes of innate behaviour, our field is limited in two respects: first, we have selected a specific object, a special group of phenomena: innate behaviour. By 'behaviour' I mean the total of movements made by the intact animal. Innate behaviour is behaviour that has not been changed by learning processes.

The question how innate behaviour can be recognized will be discussed below. Secondly, our approach is characterized by the problem previously raised: What is the causal structure underlying the observed phenomena?

Before proceeding to our task, let us consider our science with respect to other related sciences. These can be divided into (*a*) other sciences of innate behaviour, differing from ours in that they study innate behaviour from different points of view and concentrate on different problems, and (*b*) other sciences that centre around the problem of causation but do not focus on innate behaviour.

Other Sciences of Innate Behaviour

The rigid restriction to problems of causation that is advocated here is objected to by many workers in the field. They argue that we neglect two important aspects of behaviour: (1) its obvious directiveness, and (2) subjective phenomena. In my opinion, this reproach is only partly justified. The study of ethology does not intend to neglect either of these aspects in the sense of denying their importance for a general 'understanding' of behaviour. Ethologists object, however, to incorporating those aspects into a study of the causation of behaviour. I shall try to explain my point of view.

Directiveness. Many workers are impressed by the fact that behaviour, like so many other life processes, is obviously directed to a certain end (goal, or purpose). Behaviour is purposive (McDougall, 1933) or directive (Russell, 1934, 1945).

This fact, which in itself is undeniable, has a bearing on the problem of the causation of behaviour that is evaluated very differently by various workers. Many have maintained that the goal, end, or purpose of an activity serves as a causal factor controlling (in some unknown way) this activity. Russell, one of the most outspoken exponents of this view, says: 'the objective aim or "purpose" of the activity controls its detailed course . . .' (Russell, 1934, p. 12).

However, I want to emphasize from the start that this is not the view I shall take in attacking the problem of causation. Directiveness undoubtedly is a characteristic of behaviour, as it is of so many other life processes. Behaviour helps the animal to maintain itself in a hostile world, as do the functions of its intestines, its kidneys, its blood. This is the 'directiveness' of behaviour. But the statement of the fact that the heart and arteries pulsate in order to keep up a continuous flow of aerated blood to, say, the brain will not satisfy the physiologist as an answer to the question: What causes are at work in attaining this end? Neither is the statement of the fact that a falcon hunts in order to get food a satisfactory answer to the ethologist's question: What causes are at work in the falcon's body, enabling it to perform these directive activities? As will be discussed in Chapter VIII, it is of considerable interest to study directiveness, that is, to investigate to what extent and in what respects behaviour contributes to the maintenance of the individual or species. It is necessary, however, to recognize that this has nothing to do with the study of causation. There has been, and still is, a certain tendency to answer the causal question by merely pointing to the goal, end, or purpose of behaviour, or of any life process. This tendency is, in my opinion, seriously hampering the progress of ethology.

It seems to me that the main reason (though not the only one) this type of deviation from the natural course of causal study has such a tenacious hold, especially in animal ethology, is that introspection leads us to believe that our own behaviour is controlled, to a certain extent, by 'foreknowledge' of ends or goals. It is necessary to realize, however, that even in man this conclusion cannot be taken as a solution to our problem. Human behaviour, too, is dependent on physiological mechanisms, which have to be studied in order to be understood.

Subjective Phenomena. Other workers stress a different aspect of behaviour. Knowing that humans often experience intense emotions during certain phases of behaviour, and noticing that the behaviour of many animals often resembles our 'emotional' behaviour, they conclude that the animals experience emotions similar to our own. Many go even farther and maintain that emotions and other subjective phenomena are causal factors in the scientific meaning of the word. One of the most extreme exponents of this view is Bierens de Haan (1947).

Again, this is not the method we shall follow in our study of animal behaviour. Because subjective phenomena cannot be observed objectively in animals, it is idle either to claim or to deny their existence. Moreover, to ascribe a causal function to something that is not objectively observable often leads to false conclusions. It is especially dangerous in that the acceptance of the conclusion kills our urge for continued research.

To mention an instance: the conclusion that an animal hunts because it is hungry will satisfy many people at first glance. Yet the use of the word 'because' is ambiguous, since 'hungry' may be used as (1) a *convenient description* of the state of the animal, based on subjective as well as objective criteria. When the word is applied in this way, the conclusion will be clearly seen to be a very provisional one and will not satisfy the scientist who wants to know what is happening inside the animal when it is in this state. He will try to find out what impulses stimulate the muscles employed in food-seeking, where these impulses come from, and so on. But when the conclusion that the animal hunts because it is hungry is taken literally, as (2) a *causal explanation*, and when it is claimed that the subjective phenomenon of hunger is one of the causes of food-seeking behaviour, physiological and psychological thinking are confused. Although, as we said before, the ethologist does not want to deny the possible existence of subjective phenomena in animals, he claims that it is futile to present them as causes, since they cannot be observed by scientific methods.

Hunger, like anger, fear, and so forth, is a phenomenon that can be known only by introspection. When applied to another subject, especially one belonging to another species, it is merely a guess about the possible nature of the animal's subjective state. By presenting such a guess as a causal explanation, the psychologist trespasses on the domain of physiology.

Therefore, though I do not want to belittle the importance of a study of either the directiveness of behaviour or of the subjective phenomena accompanying our and possibly the animals' behaviour, I want to stress the paramount importance of recognizing the limited nature of such a study, as well as the limited nature of the study of causation. The study of directiveness, the study of subjective phenomena, and the study of causation are three ways of thinking about behaviour, each of which is consistent in the application of its own methods. However, when they trespass into each other's fields, confusion results.

Other Causal Sciences of Behaviour

Much more numerous than the workers who study behaviour from another point of view than ours are those who formulate their problems in causal terms but concentrate on more or less different phenomena.

Physiology. A first category is formed by those usually called 'physiologists'. Among them many specialize in the physiological study of one or more sense organs; others concentrate on muscle or on nerve physiology. In a sense they all study the physiology of movement, and their results are of great importance to the student of behaviour. However, knowledge of the processes taking place in the sense organs, in the nerve-fibres, and in the muscle-cells is not quite sufficient for an understanding of animal movement. As will be shown later, behaviour is always the outcome of a highly complex integration of muscle contractions. Although some neurophysiologists study integrative processes, their work is usually confined to the relatively simplest types of integrative activity, namely, the various types of reflex movements. It will be one of the tasks of this work to show that even relatively simple instinctive activities are of a much more complex nature, that they belong to a higher integrative level, and that it is our job to carry our analysis from these high levels down to the level of the neurophysiologist, the sense physiologist, and the muscle physiologist.

It should be stressed that although the 'objective' student of behaviour differs from these physiologists in the phenomenon which he studies, both they and he are tackling fundamentally the same problem. Therefore, the field covered in the first five chapters of this book could be called the 'physiology of behaviour'. Mainly for historical reasons the concept of physiology has been confined to a study of the functions of separate organs. The more synthetic tendencies of our time tend to break down the more or less artificial barriers between separate organs and to direct attention to their integrated working. While in itself it is perhaps unimportant whether or not the causal study of behaviour is considered part of physiology, it certainly is important to be aware of the fact that the causal study of behaviour uses the same methods that are employed in the causal study of the functions of the three organ systems that play a part in behaviour. For it is this identity of method that will enable us, in the long run, to combine all these fields into one 'physiology of movement'.

Psychology. A second category of students of behaviour, the psychologists, differ from the physiologists and resemble the ethologists in that they study behaviour as a whole. But both in regard to the phenomena studied and the problems that are put in the foreground, we must distinguish between two kinds of psychologists.

In America most psychologists apply the 'behaviouristic' method, which is fundamentally just another name for the 'objective' method. They do not, however, concentrate on innate behaviour; rather they specialize in the higher types of behaviour. This specialization has developed as a consequence of (1) the general interest in man and human conduct, and (2) the general acceptance of man's evolutionary descent from ape-like ancestors. These factors have brought about a preoccupation with what is often called 'prehuman' behaviour in mammals. The result has been a certain neglect of innate behaviour, which has led in some instances to entirely unwarranted generalizations. In my opinion, this disregard of innate behaviour is due to the fact that it is not generally understood that learning and many other higher processes are secondary modifications of innate mechanisms, and that therefore a study of learning processes has to be preceded by a study of the innate foundations of behaviour.

While American psychologists on the whole apply objective methods, this is by no means true of their European colleagues. It would carry us too far, and it would, moreover, be beyond my power to make a critical inventory of the multitude of psychologies in existence. For our present purpose it may suffice to point out that while the phenomena studied in this book are for the most part the same phenomena as those studied in psychology, the type of approach will be physiological or objective.

SOME CHARACTERISTICS OF THE OBJECTIVE STUDY OF INSTINCT

Description

As in every field of science, the description of the observed phenomena, which has to precede their causal study, is not random but selective; that is to say, it is adapted to the problem under consideration. It is, therefore, necessary to point to some characteristics of the descriptions required for our study.

1. Because it is our task to analyse behaviour as co-ordinated muscle activity, the ultimate aim of our description must be an accurate picture of the patterns of muscle action. Except in some especially simple cases, this has never been done, probably because most workers are only dimly aware of the necessity.

FIG. 2. Herring gull chick begging for food.

The preparation of such a description is an exacting task. It requires an objective recording of movements, in which the help of still and motion pictures is often indispensable, and a study of the muscles taking part in each movement. As an instance of a step in this direction, Baerends and Baerends's study (1950) of the behaviour of cichlid fish may be mentioned.

2. For some purposes, however, the descriptions of behaviour need not be complete but merely sufficiently accurate to characterize the behaviour. An activity is described in this way when the causal factors responsible for a certain reaction as a whole are being studied. For instance, in an investigation of the external stimuli releasing the pecking response of a herring gull chick (Fig. 2), it is sufficient to characterize the pecking reaction by describing only some of its features. However, as soon as this reaction has to be broken down into its components, the description has to be much more detailed.

3. Whenever an activity of one animal causes a response in another, the description of this activity has to fulfil still different requirements: it will be concentrated on those peculiarities that influence the other animal's response.

Thus the descriptive part of an ethological study may vary with the purpose it is serving. However, this is still sadly neglected in many papers appearing at present. Too many descriptions are rather haphazard collections of fragmentary details.

Ethograms

Special emphasis should be placed on the importance of a complete inventory of the behaviour patterns of a species. It is a natural tendency of the experimental worker to select a special problem, as, for instance, colour vision, or homing, or the delayed response. This specialization is often accompanied by a narrow point of view and a neglect of other aspects of behaviour. The resulting generalizations, based on too limited a foundation, may give rise to sterile controversies.

Because the importance of extensive descriptive work as a necessary preparation for experimental work is not generally realized, I shall give some instances of these controversies.

Discrimination of Colour by Honey Bees. In order to study the ability of honey bees to distinguish between different colours, von Hess (1913) confronted honey bees in a dark room with two lights of different colour and intensity. He found that the bees under these conditions moved to the brighter of the two lights, irrespective of their colour. For example, when confronted with red and green, the bee could be forced to choose either one of the colours by changing the intensities. An apparently selective reaction to colour thus proved to be a reaction to intensity. The natural conclusion would seem to be that a bee is colour-blind.

Von Frisch (1914), impressed by the ability of the honey bee to select blue and yellow flowers amidst a multitude of other colours and hues, presented bees with yellow and blue papers together with a variety of grey papers of many shades from white to black. He found that the bees reacted selectively to the coloured papers. His experiments included a number of controls and are no less reliable than those of von Hess. Therefore, von Frisch concluded that bees could distinguish between colours, and a lively discussion between the two workers followed.

The more recent view, derived from a consideration of similar reactions in many animals, is that a species may be able to distinguish between colours and, in a way, be colour-blind as well. For we know now that innate reactions show a selective responsiveness to very special stimulating situations. Different reactions of one and the same individual are, as a rule, dependent on very different releasing external stimuli. Thus one reaction may be released by light of a certain wave-length, while another reaction responds to intensity, independent of wave-length. Anybody studying the first reaction would come to the conclusion that the animal can see colours, whereas a worker who confines himself to the study of the second reaction would judge the animal colour-blind.

Knowledge of the whole behaviour pattern helps us to recognize the relative value of each one of these conclusions and prevents us from describing as incompatible the conclusions drawn from the study of what prove to be two different sorts of behaviour.

Delayed Reaction. My next example of how important it is to know the entire behaviour pattern of a species before starting experimental work concerns the so-called 'delayed reaction'.

The 'delayed response' (Carr-Hunter) was originally used to test an animal's ability to retain an 'idea'. It is now considered to be one of a number of methods to study memory. The experimental set-up is nearly always the same. In a multiple-choice apparatus one door is indicated by a stimulus (either conditioned or unconditioned). The animal is not allowed to react while the stimulus is present but only a certain time after the stimulus has disappeared. The maximum delay allowed by the animal is a measure of its memory.

The value of the method has often been doubted, for instance by Maier (Maier and Scheirla, 1935), who pointed out the remarkable dis-

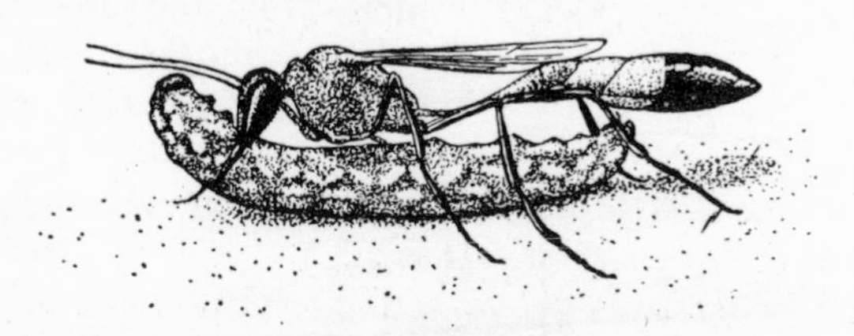

FIG. 3. The digger wasp *Ammophila campestris* with prey. After Baerends, 1941.

parity between the values found for the gorilla (48 hours) and for the orang utan (5 minutes).

Criticism of a more fundamental nature has been given by Baerends (1941) on the basis of his study of the digger wasp, *Ammophila campestris* (Fig. 3). A female of this species, when about to lay an egg, digs a hole, kills or paralyses a caterpillar, and carries it to the hole, where she stows it away after having deposited an egg on it (phase *a*). This done, she digs another hole, in which an egg is laid on a new caterpillar. In the meantime, the first egg has hatched and the larva has begun to consume its store of food. The mother wasp now turns her attention again to the first hole (phase *b*), to which she brings some more moth larvae; then she does the same in the second hole. She returns to the first hole for the third time to bring a final batch of six or seven caterpillars (phase *c*), after which she closes the hole and leaves it for ever. In this way she works in turn at two or even three holes, each in a different phase of development (Fig. 4). Baerends investigated the means by which the wasp brought the right amount of food to each hole. He found that the wasp visited all the holes each morning before leaving for the hunting-grounds. By changing the contents of the hole and watching the subsequent behaviour of the wasp, he found that (1) by robbing a hole he could force the wasp to bring far more food than usual; and (2) by adding larvae to the hole's contents he could force her to bring less food than usual. But these changes influenced the wasp only when they were made before the first visit. Any later change had not the slightest effect. The situation in the hole at the time of the first visit determined the wasp's behaviour for the whole day. On the basis of the situation she found in the three holes she was attending, the wasp chose the one she supplied during the rest of the day. In this way Baerends proved that the response of the wasp could be delayed as long as fifteen hours.

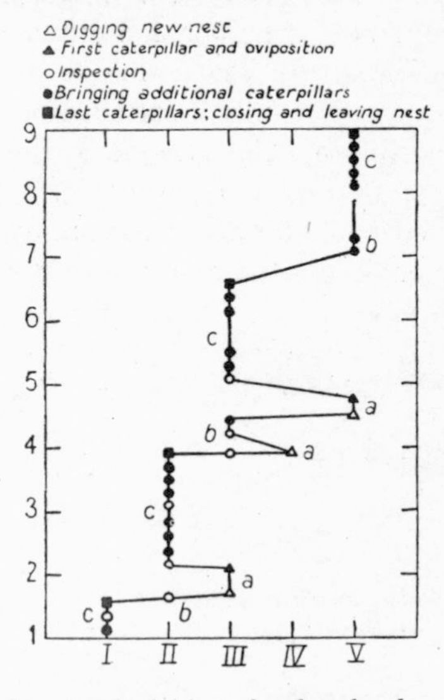

FIG. 4. Activities of a female *Ammophila campestris* at five successive nests (I–V) in the course of eight days (1–8). *a*, *b*, *c*, the three phases of nest-care. After Baerends, 1941.

Apart from the amazing length of the delay, this fact demonstrates the importance of carefully selecting that part of normal behaviour in which a delayed reaction normally plays a part. It is highly probable that a delayed-reaction test carried out in the conventional way, for example with a foraging wasp, would not have the slightest result. Thanks to the preparatory survey of the whole behaviour pattern of the species, Baerends was able to find out where a delayed reaction played a part. This peculiar 'localization' of higher processes to certain phases of behaviour is by no means rare. It will be discussed more fully in Chapter VI.

The Need of a Broad Basis for Generalizations

Innate behaviour is found in so many and in such widely differing species of animals that it is clearly undesirable to confine our study to a too limited number of species. The concentration of American psychologists on problems of 'prehuman' behaviour has led them to select mammals, and even a very small number of species, as their special objects. It must be realized, however, that mammals comprise, in comparison with the rest of the animal kingdom, a very small group. Moreover, they are specialized for intelligent behaviour in its widest sense—for what could be called 'cortical' processes. A study of instinctive behaviour, which is of so much more widespread occurrence, naturally has to be based on a broader foundation lest the generalizations run the risk of being premature. To mention an instance, E. C. Tolman's book *Purposive Behavior in Animals and Men* (1932) is dedicated to *M(us) N(orvegicus) A(lbinus)*, 'where perhaps,' the writer states in his preface, 'most of all, the final credit or discredit [of his book] belongs.'

In spite of the high respect deserved by the interesting work done with rats, one should be a little sceptical of the laboratory rat as a representative of the whole animal kingdom.

Another undue generalization based on too narrow a foundation was Loeb's theory of tropisms, which gave rise to an unnecessary controversy (see Loeb, 1918). Loeb, realizing the great difference between a reflex movement and purposive or directed behaviour, laid great stress on his discovery of tropisms. Various animals (*Planaria*, *Daphnia*) that normally move straight to a light could be forced, by unilateral blinding, to move in circles towards the side of the normal eye. From this and other experiments (see Chapter IV) he concluded that equal stimulation of right and left receptors was necessary to balance the muscular contractions of the right and the left side. In this way he explained a simple oriented movement as a co-ordination of undirected components, or reflexes. He demonstrated this principle by the construction of an 'artificial heliotropic machine'.

From this in itself important fact he drew two more general conclusions, both of which were, as will be shown below, unwarranted: (1) all oriented movements are tropisms or combinations of them; (2) all behaviour is composed of tropisms. However, subsequent research by many others, both in America and Europe, has shown that there are other types of oriented behaviour.

Jennings (1923), for instance, laid great stress on the principle of trial-and-error orientation. Furthermore, there are genuinely directed activities that are not dependent on balanced stimulation of two receptors. For example, a dragonfly capturing an insect is not directed to its prey by a tropism, for its directed flight is not impaired when one eye is blinded. Kühn (1919), in a brilliant paper that is not sufficiently known in the English-speaking world, showed that there are at least three totally different types of orientation movements, of which the tropism, or, to use Kühn's terminology, the tropotaxis, is only one. Different species differ in the orientation mechanisms they have at their disposal, and it has even been shown that each species possesses a relatively great number of them. For instance, the grayling butterfly (*Eumenis semele*) (Fig. 5) escapes from a predator by flying towards the sun. This is a tropotaxis, as can be shown by unilateral blinding, which causes circular motion. The males of this species have another directed reaction: they follow passing females. This reaction is entirely dependent on visual stimuli too. However, unilaterally blinded males, while showing circular motion when escaping, will follow females in a straight course immediately afterwards (Tinbergen, Boerema, Meeuse, and Varossieau, 1942).

Here again, then, undue generalization led to a controversy between Loeb's and Jennings's views, but further studies, based on broader foundations, showed that both exponents were partly right and that the principles discovered by them (as well as other principles that they did not discover) each played a part.

Experimental Procedure

The foregoing considerations concerning the need for a broad descriptive approach as a preparation to experimental work profoundly influence our evaluation of the experimental method. Every experimental worker has a certain inclination to stick to a method that has served him well and has given him results. This is not unexpectedly so in cases where elaborate experimental set-ups have been made use of. However, this attitude often leads to a certain exaggeration of the value of a special type of experiment. In American psychology, for instance, the effect has been to centre the whole field around a few techniques, such as mazes, problem boxes, multiple-choice apparatus, and so forth. However valuable these methods have been and still are, their value compared with that of thousands of other methods (already applied or imaginable) is rated much too high. The scientist who limits himself in this way becomes more and more inclined to let his work be determined by a certain preoccupation with the question: 'What can I do with this method?'

This, however, is not the natural way of expanding a science. I want to stress the necessity of a return to a sounder approach, in which the problem is the primary concern and dictates the special observational and experimental procedures required for its study. In the long run, it is more useful to ask, 'What method shall I apply to solve this problem?' rather than, 'What can I do with this method?'

The amazing achievement of *Ammophila campestris* in delaying a reaction throws another sidelight on the problem of experimental procedure: it shows that 'standardization' of method has to be based on a knowledge of the behaviour pattern as a whole. To put different species in exactly the same experimental arrangement is an anthropomorphic kind of standardization. As I have already said, there is little doubt that *A. campestris* would fail entirely in the conventional delayed-reaction test. In view of the differences between any one species and another, the only thing that can be said for certain is that one should *not* use identical experimental techniques to compare two species, because they would almost certainly not be the same to *them.*

RELATIONS TO OTHER FIELDS OF ZOOLOGY

As we have seen, the 'objective' study of innate behaviour is closely related to, or rather is part of, physiology, having affinities to neuro-

FIG. 5. The Grayling butterfly (*Eumenis semele*) in natural surroundings. After Tinbergen, Meeuse, Boerema and Varossieau, 1942

physiology, sensory and muscle physiology, and endocrinology. It is worth while considering briefly the relations with other fields of zoology as well.

Ecology

Relations between the animal and its environment are always mutual. Each species, however low in the systematic scale, is continuously protecting itself against the destructive forces of the environment. Behaviour is an essential element in the equipment serving this end. In fact, the closer our study of behaviour becomes the more we are convinced that numerous details of behaviour are 'adaptive', that is to say they play a part in the relations between the animal and its environment. Although this is self-evident to every ecologist acquainted with such phenomena as, for example, habitat selection, host selection, reactions to predators, food preferences, &c., only few workers recognize the amazingly high degree of adaptiveness to be found in numerous behavioural characteristics.

Sociology

Animal aggregation and co-ordination can only be fully understood by a study of the social behaviour of animals, for animals are rarely forced to congregate entirely passively, like dust grains in a whirlwind. Rather, they come together as a result of special behaviour controlled by specific sensory stimuli, enabling them to distinguish and select members of their own species.

Also, the manifold forms of co-ordination between individuals, towards which congregation usually is but the first step, are based upon highly specialized behaviour patterns. This is why ethology is, perhaps, the most important adjunct of sociology.

Taxonomy

It is only for practical reasons that taxonomy is based primarily on morphological characters; while these can be studied in the museum and are relatively easy to detect, ethological characteristics can only be seen in the live animal under natural conditions and are not displayed continuously. However, this does not mean that species do not differ ethologically nor even that behavioural characteristics are always of minor importance to the taxonomist. In 'difficult' groups, certain behaviour elements may be used as species characters where morphological study fails, or at least presents difficulties. As Lorenz (1937*a*) points out, the family of the Columbidae (pigeons) cannot be characterized by any group of morphological characters distinctive of all species of the family, whereas there is one behaviour element which is present in every species: unlike other birds they all make sucking motions while drinking. Only the related family of the Pteroclidae (sand grouse) have the same way of drinking. Whitman's studies of pigeons (1919), Heinroth's work on numerous birds, especially ducks (1911), Lorenz's paper on surface-feeding ducks (1941), Delacour's and Mayr's treatise on ducks and geese (1945) and several other works demonstrate the usefulness of ethological characters for taxonomic purposes. Also, the so-called sibling species (Mayr, 1942), while being very similar morphologically, may be very different ethologically: two sibling species of digger wasps, originally considered one species, have recently been recognized as two distinct species by a student of their behaviour (Adriaanse, 1947).

The selection of behavioural elements useful for taxonomic work has been greatly facilitated by the work of Lorenz. Though Whitman was the first to point out the remarkable stereotypy of certain movements in birds, it was Lorenz who first characterized this type of movements (the 'fixed patterns') ethologically and physiologically, and showed that, like morphological elements, they are homologous in related species. This aspect of ethology will be discussed more fully later (Chapters III, V, and VIII).

Evolution

There are two ways in which the student of behaviour comes into contact with the central problem of biology: that of evolution. First, as will be clear from the last paragraph, the study of homologies inevitably leads to a dynamic, historic interpretation of the divergences of homologous elements found in related species and of the convergences among non-homologous elements in different groups. Although much fragmentary work has been done in this field and scattered facts are to be found in the literature, Whitman, Heinroth, and Lorenz are the only authors who consciously made systematic attempts in this direction. Recently Mayr and Spieth have started a promising programme along these lines in *Drosophila* (see Spieth, 1947).

A second way in which the study of evolution makes contact with the study of behaviour has been pointed out by Mayr (1942). The paramount importance of the study of isolating mechanisms preventing hybridization between closely related species is obvious when one considers that speciation would not be possible in groups without such isolating mechanisms.

In the past, relatively too much stress has been laid on morphological isolating mechanisms. In by far the majority of cases animals actively select their mates and the decisive isolating mechanisms are of an ethological nature. Again, this field has not been developed systematically, but there is enough fragmentary evidence on the existence of such ethological isolating mechanisms.

II

BEHAVIOUR AS A REACTION TO EXTERNAL STIMULI

'SPONTANEOUS' AND 'REACTIVE' BEHAVIOUR

ONE of the old controversies in the study of animal behaviour concerns the question of whether behaviour is 'spontaneous' or whether it could be explained as a combination of simple reactions to the environment. Most workers of physiological, objective temper have claimed that behaviour was all 'reaction'. This attitude was natural in so far as the discovery of the simple reflex movement made it possible for the first time to study, by physiological methods, a type of co-ordinated functioning of the three organ systems involved in behaviour. The early development of 'reflexology' and, later, the discovery of the 'conditioned reflex', caused a wave of optimism in physiological circles, and several prominent physiologists claimed that reflexes and conditioned reflexes were the only elements of behaviour. Pavlov simply identified 'instinct' with 'reflex' and stated, for instance, that the tendency to collect money in man is 'an Instinct or Reflex'. Science, according to Pavlov, is the result of the activation of the 'What-is-that-Reflex', and so on (see Pavlov, 1926). Loeb's theory of tropisms is another example of this generalization of reflexology.

Spontaneity, on the other hand, has always been stressed by psychologists. Many of these psychologists were definitely superior to the reflexologists in their knowledge of animal behaviour as a whole. But, unfortunately, many of them had a certain disinclination to objective study, and this has created considerable confusion and delay in the development of our science, because it has helped to establish the opinion that spontaneity is not susceptible of objective study. Somehow it was assumed that, once it could be shown that a certain type of behaviour was 'spontaneous' (that is, independent of external stimulation), it would be futile to attack it with physiological methods.

We are at present in a position to say that both opinions contain part of the truth. Behaviour is reaction in so far as it is, to a certain extent, dependent on external stimulation. It is spontaneous in so far as it is also dependent on internal causal factors, or motivational factors, responsible for the activation of an urge or drive.

These two types of causal factors can both be studied by objective methods, although they require entirely different techniques. I shall discuss them both; in this chapter I shall first confine myself to the study of external stimuli.

This study has to be made in two steps. First we want to know what stimuli the sense organs of a given animal can receive. Second, the actually effective stimuli, those responsible for the release of each reaction, have to be determined.

THE POTENTIAL CAPACITIES OF THE SENSE ORGANS

One of the first things that impresses the student of animal behaviour is the fact that the working of the animal's sense organs is not the same as ours. For instance, many animals, like starfish, snails, flies, are completely deaf. Others are blind, or nearly so, &c. But, on the other hand, some animals are able to hear sounds that are inaudible to us (locusts, bats) or may smell odours that are entirely imperceptible to us (many mammals, moths). A careful study of sensory capacities reveals the fact that almost no two species have exactly the same capacities. Von Uexkull (1921) has emphasized this by saying that each animal has its own *Merkwelt* (perceptual world) and that this world is different from its environment as we perceive it, that is to say, from our own *Merkwelt*.

The first task, therefore, when tackling the study of a new species is a careful examination of the capacities of its sense organs. The classical method used for this purpose is the conditioning method, which has been developed admirably by von Frisch and his school. Von Frisch reasoned that if an animal's sense organs are affected by a change in the environment the animal can be conditioned to show a response to it. The general procedure is the following. A certain reaction, for instance escape, or feeding, is conditioned to a definite, simple change in the environment, for instance to the experimenter blowing a whistle. When this has been accomplished, the next step is to ascertain to which sensory modality the conditioned stimulus belongs. In the present case, if a response to sound is wanted, one has to be sure that it was actually the sound the animal was reacting to and not, for instance, the experimenter's movement in bringing the whistle to his mouth. This is done by comparing the reaction to the complete situation (blowing a whistle) with that to the visual part of it, or rather to the complete situation minus the auditory part. This is done by bringing the whistle to the mouth but not blowing it. If the animal reacts exclusively to the first situation it is obvious that the response was auditory. Thus, by systematically probing the animal's potential reactivity to many different environmental influences, a survey of its sensory capacities can be made.

The aim of the ethologist in doing this type of work is slightly different from that of the sense-physiologist. While the latter makes a survey as a preliminary to studies of the physiological mechanisms underlying sensory reception, the former's main interest is to know which properties of the external world can influence behaviour and, equally important, which properties cannot.

The two methods usually supplement each other. Thus the functions of the lateral line organs of fish have been studied by Dijkgraaf (1934) and by Sand (1937). Sand, working with rays, used the electrophysiological method, registering action potentials in the sensory nerve.

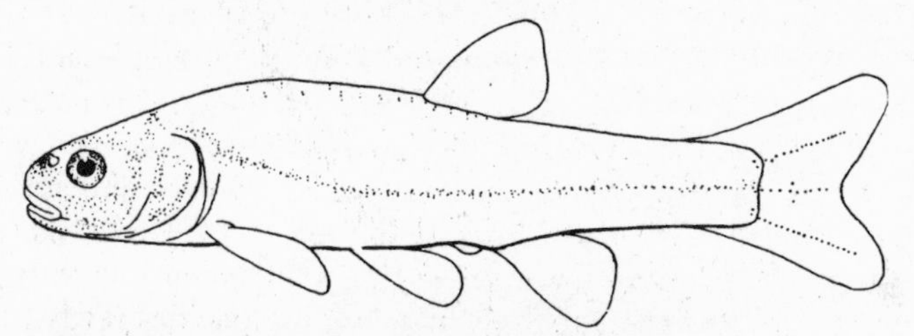

FIG. 6. Lateral line organs of the minnow (*Phoxinus laevis*). After Dijkgraaf, 1934.

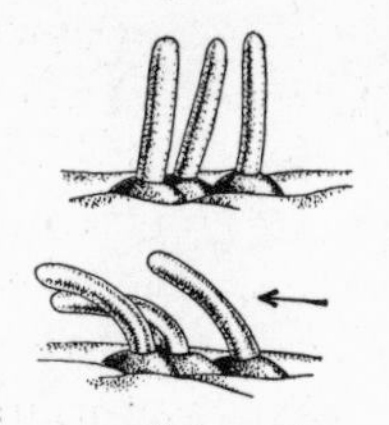

FIG. 7. Lateral line organs of the minnow, at rest (above) and stimulated by a water current (below). After Dijkgraaf, 1934.

Dijkgraaf trained various species of teleost fish, particularly minnows, to respond by special behaviour patterns. Both authors found that the lateral line organs are sensitive to mechanical stimuli, viz., local water movements (Figs. 6, 7). Dijkgraaf then proceeded to study the part played by such stimuli in the normal life of fish and found that they helped them locate prey, and predators. Sand's further work concentrated on the mechanism of sensory stimulation.

The reactions of honey bees to polarized light provide another example of this diversity of aims. During his studies of the 'language' of honey bees von Frisch (1949, 1950) found that bees could orientate themselves to the sun even if the sun itself was invisible to them. Extensive tests in which the plane of polarization of the light was changed by means of a polaroid sheet proved that the bees reacted to the pattern of polarization of the sky.

Autrum (1950), registering action potentials from the optic nerve in response to illumination of one or at least very few ommatidia, added to our understanding of the underlying sensory mechanism by showing that one ommatidium can distinguish polarized from ordinary light of the same intensity.

Although an exhaustive review of the wide array of facts that have been brought to light in this field would be out of place, it is necessary to consider some of the general problems a little more closely.

Sensitivity

First of all we want to know the limits of sensitivity of the sense organs. There are limits of two kinds: of intensity and of quality. Visual receptors have a threshold of intensity below which light is an ineffective stimulus. Systematic studies of this problem are rare. Owls definitely have a lower threshold than man; also, there are differences between species of owls that are correlated with differences in their habits. Thus under experimental conditions guaranteeing visual orientation the barred owl (*Strix varia*) is able to pounce directly on the prey from a distance of 6 feet, when the light intensity falling on the floor was calculated to be 0·000,000,73 foot-candle, which is between one-hundredth and one-tenth of the light intensity that man requires for vision. A less nocturnal species, the burrowing owl (*Speotyto cunicularia hypohaea*), has about the same ability to see in weak light as has man (Dice, 1945).

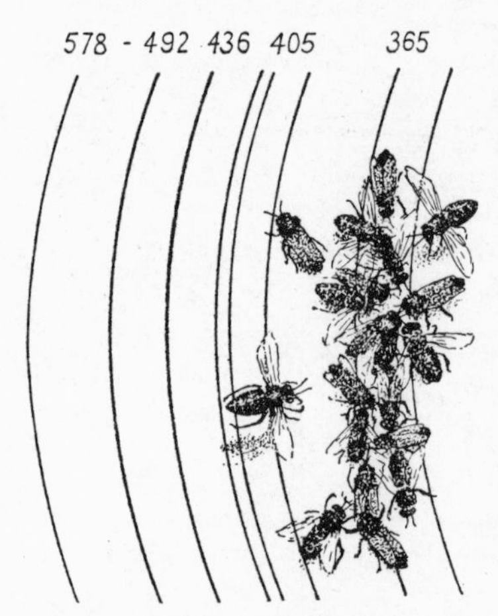

Fig. 8. Honey bees reacting to the ultra-violet part of the spectrum. After Kühn, 1928.

The limits of colour sensitivity of light receptors have been studied in more detail. It is now known that the eyes of various animals may be sensitive to other parts of the spectrum than is the human eye. Thus honey bees, like many other insects, are less sensitive to light of long wave-length than is man (von Frisch, 1914; Kühn, 1927). The same, though to a lesser extent, is true of the little owl (*Athene noctua vidalii*) (Meyknecht, 1941). On the other hand, honey bees are able to respond to ultra-violet light down to at least 3,500 Å (Kühn, 1927) (Fig. 8), their reactions to the flowers of *Papaver rhoeas*, which are of a very pure red to the human eye, are entirely guided by the ultra-violet light it reflects (Lotmar, 1933).

Whether any animal is able to see infra-red light is still doubtful. Vanderplank's (1934) results with tawny owls (*Strix aluco*) have not been corroborated and later experiments by Hecht and Pirenne (1940) with *Otuis asio*, and by Meyknecht (1941) with *Athene noctua* render it highly improbable that owls are sensitive to infra-red 'light'.

The same problems of intensity and quality limits are encountered in other sense organs. Studies on the lower limit of sound intensity are also rare. Räber (1949) and Dice (1947) gave some remarkable facts on the amazing sensitivity of owls, facts that are not surprising in view of the specialized anatomy of their ears.

Limits of pitch sensitivity are not the same for every species. Von Frisch's and Stetter's experiments (1932) with fish revealed an upper limit in minnows of about 5,000–7,000 cycles per second, in *Amiurus* of about 13,000 cycles, which is slightly below the upper limit found in man (see also von Frisch, 1938). A very high upper limit has been found in bats. The supersonic cries of bats, serving the purpose of 'echolocation' (Fig. 9) have a frequency of about 50,000 cycles per second (Griffin *et al.*, summary in Dijkgraaf, 1946); in preliminary tests carried out by Dijkgraaf (1946) it was shown that sounds at least as high as 40,000 cycles per second can be received by bats.

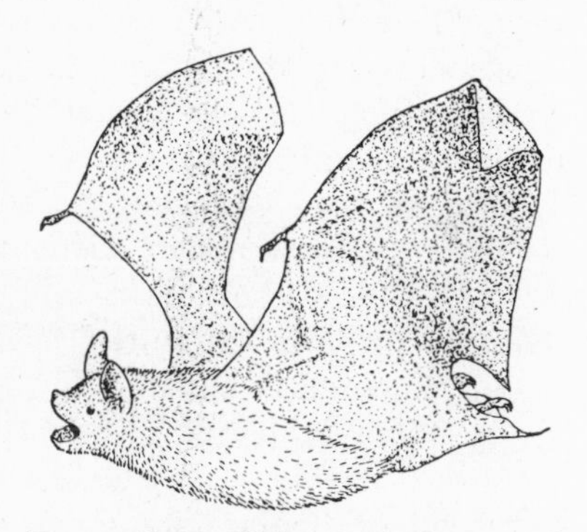

FIG. 9. Bat in flight, producing sound for the purpose of echolocation. After Freeman.

In chemoreception the same problems of sensitivity are encountered. Although quantitative data are rare, it is known, for example, that macrosmatic mammals have a much lower threshold than man. Matthes found (1932*a*, *b*) that in *Cavia* the threshold for nitrobenzol and bromostyrol was about one-thousandth of that of man.

Qualitative differences of sensitivity between species are more striking in the case of chemoreception than in other sensory fields. Von Frisch (1934) tried to find out systematically what substances are taken by bees as substitutes for sugar. He found that on the whole these substances were the same as those having a sweet taste for man; however, some compounds that were accepted by bees are tasteless to man and some compounds that are considered sweet by man were not accepted by the bees.

Many fish respond with escape reactions to a substance which dissolves into the water when the skin of a fish of the same species is damaged by a predator. This is an olfactory stimulus, that is a stimulus received by the sensory cells in the nasal cavity. The substance is not volatile, and therefore could not give off a scent for land animals (von Frisch, 1942).

Discrimination

The ability to distinguish between different stimuli belonging to one sensory modality offers another problem. Again, intensity and quality have to be considered separately.

In animals with well-developed eyes the discrimination of intensities determines the richness in hues of the visual field, or the 'gradation' as it is called in photography. In species with a diffuse photoreceptor the problem is one of discriminating between differences over a period of time. Thus many marine molluscs, worms, and cirripedes react by escape movements of various kinds to a decrease in light intensity ('shadow-reflex') which acts as a sign stimulus indicating the presence of a predator. In the crustacean *Balanus* the minimal stimulus is a darkening of 5 per cent. (von Buddenbrock, 1931).

Other species react to an increase in light intensity. The mollusc *Mya arenaria* withdraws its siphon when the illumination is increased and thus prevents exposure (Hecht, 1934).

Qualitative discrimination in the visual field has been studied extensively. Most studies in this field are confined to the question of whether an animal is able to see colours, and go no farther than to say it can, i.e. is able to distinguish four or five different colours from each other and from any mixture. However, in some cases the degree of discriminative power has been studied systematically. Thus Grether (1939) found that various Old World monkeys had about the same ability to discriminate between hues as man, distinguishing in the red region between two lights differing by about 10 mμ in wave-length and in the blue-green by about 9 mμ.

The colour vision of honey bees extends into the ultra-violet which is distinguished as one or more separate qualities from other colours (Lotmar, 1933). Whereas the first results concerning the colour vision of insects (von Frisch, 1914; Knoll, 1921–6) gave the impression that they were only able to distinguish two groups of colour, the yellow group and the blue-violet-purple group, it was afterwards found that at least honey bees were able to distinguish between many colours within each group (Lotmar, 1933).

It is still too early to draw general conclusions as to which animals can distinguish between colours and which cannot. However, it seems that, whenever the colour vision in a species is accurately tested, an ability to distinguish between colours is found. In vertebrates, only the results obtained with some mammals give rise to doubt, though colour vision has been proved for the hedgehog (Herter, 1933), the squirrel (Locher, 1933) and several primates (Grether, 1930). Negative results may be due to the fact that many animals react to special colours only in special circumstances. (For an explanation of this phenomenon, see below, p. 25.)

In the auditory field, accurate studies of discrimination of sound of different pitch have been published by Stetter (1933) and by Wohlfahrt (1932). It was found that minnows are able to distinguish an interval of even a half-tone.

Studies of chemical discrimination have, for instance, been carried

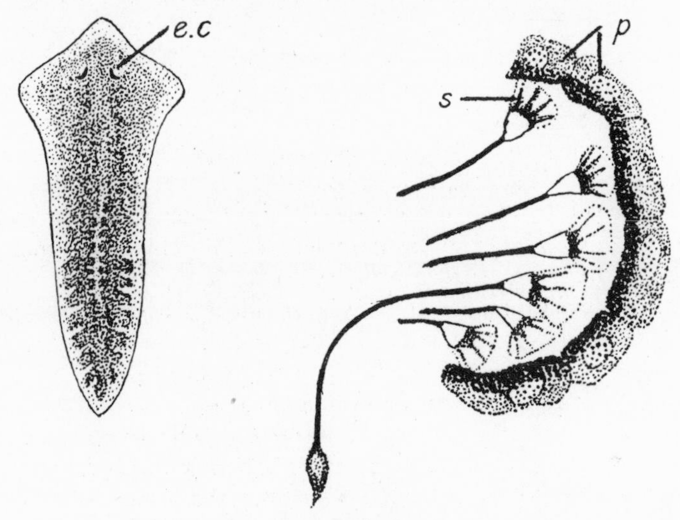

FIG. 10. *Planaria*, and one of its eye cups enlarged. After Taliaferro, from Kühn, *Grundriss der allgemeinen Zoologie*, 8th ed., Georg Thieme, Leipzig. *e.c.*, eye cup *s.*, sensory cell; *p.*, pigment.

out with dogs. In distinguishing between scents and in their ability to recognize a particular scent among a mixture of others, dogs are far superior to man.

Localization

The ability to localize the source of a stimulus in space is of great importance and it is because of their power of spatial localization that we may distinguish 'higher' from 'lower' sense organs. Localization has two aspects: direction and distance.

Localization of direction is developed to the highest degree in the eye. From an entirely diffuse sensitivity to light as found in many lower organisms numerous evolutionary lines have led to the development of more or less specialized eyes. *Euglena* offers a simple type (Mast, 1911), *Planaria* a more advanced type (see, e.g., Kühn, 1939) (Fig. 10). The eye cups of many worms and molluscs, reaching a high degree of perfection in nautiloids (Fig. 11), are a further step towards analysing the visual field. Their coming together to form compound eyes is a further specialization reaching its highest development in insects (Fig. 12); the development of a lens within one large cup, such as has taken place in molluscs and vertebrates, is a specialization in another direction. Of

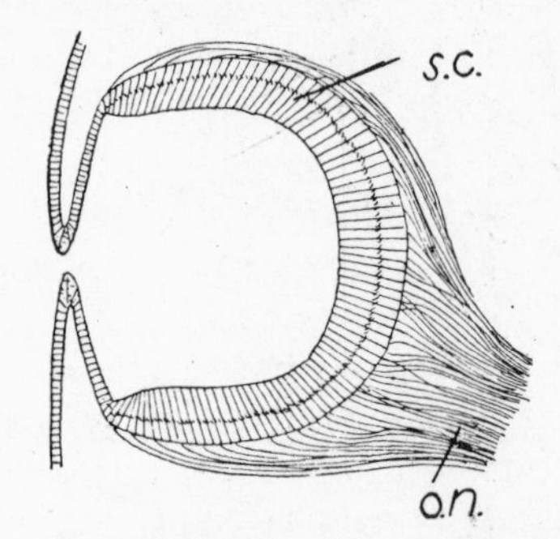

FIG. 11. Eye of *Nautilus*. After Kühn, *Grundriss der allgemeinen Zoologie*, 8th ed., Georg Thieme, Leipzig. *s.c.*, sensory cells; *o.n.*, optic nerve.

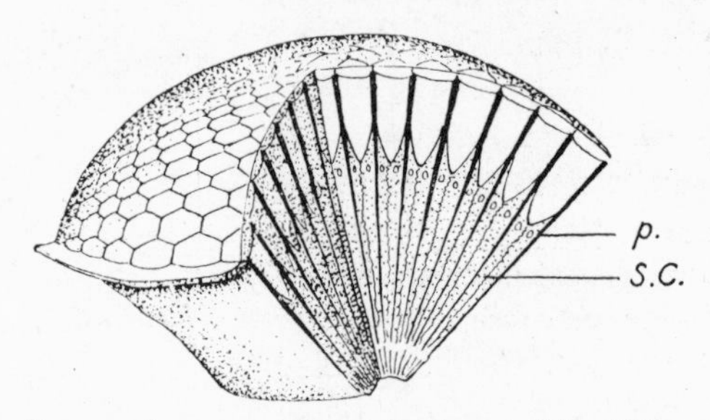

FIG. 12. A compound eye of an insect. After Kühn, *Grundriss der allgemeinen Zoologie*, 8th ed., Georg Thieme, Leipzig. *p.*, pigment; *s.c.*, sensory cells.

these two types, the latter has led to the most successful apparatus with respect to discrimination of direction. This power of discrimination of direction, or visual acuity, is probably best in birds, where preliminary studies revealed a minimum visual angle of 21° (Schuyl, Tinbergen, and Tinbergen, 1936). The *minimum visible* of the compound eye of a honey bee has been found to be about 1° (Hecht and Wolf, 1929); it is not probable that other insects surpass this value substantially.

Sound localization is, on the whole, much less accurate. Experiments have been carried out with dogs by Engelmann (1928), who found that their sound localization was about twice as accurate as that of man.

Whereas in vertebrates, sound localization depends on the co-operation of two ears, the results obtained by Autrum (1940) with locusts seem to indicate that localization of direction can be effected by each of the two tympanal organs separately.

The chemical senses contribute much less to localization of direction. Whereas all chemical sense organs dependent on contact stimulation, like taste, do not allow the localization of an object at even a short distance, the sense of smell may enable an animal to localize an odorous object. In most species, as among mammals, this may be accomplished by combining the chemical sensory data with those supplied by movements of the medium (wind). However, some animals, usually such as live in stagnant water, are able to draw a water current to them, and thus take samples from various directions. The marine snail *Buccinum undatum* takes its samples by aiming its siphon and sucking in a narrow current of water. This current is directed against a field of chemoreceptive cells and in this way *Buccinum* is able, by taking successive samples from different directions, to 'dissect' the chemical surroundings and perceive a spatial chemical pattern (Brock, 1936) (Fig. 13). The distance from which *Buccinum* is able to get its samples does not exceed a few centimetres. Some crustaceans, using the same method in principle, succeed in taking samples from a distance of about 15 cm. (Brock, 1926).

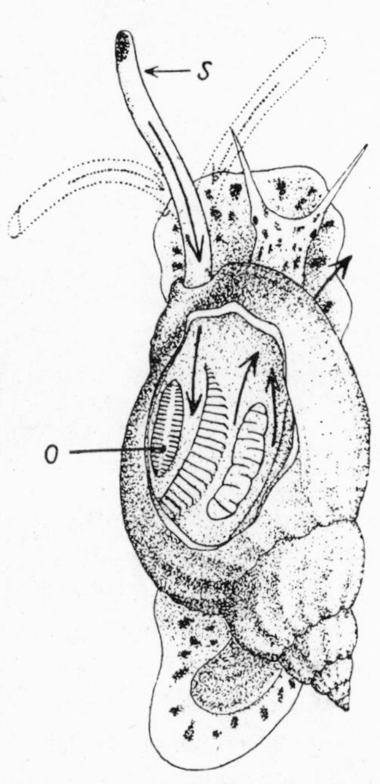

FIG. 13. A whelk (*Buccinum undatum*) taking water samples. *S.*, siphon; *O.*, chemoreceptor (osphradium). After Brock, 1936.

In the water bug, *Notonecta glauca*, location of direction is possible with the aid of touch receptors. This species reacts to minute surface ripples and very accurately locates the direction of a moving prey at distances up to 15 cm. (Baerends, 1939) (Fig. 14).

The ability to judge distance, and thus to build up a three-dimensional picture of the environment, depends almost exclusively on the eye. One of the principles involved is binocular vision. A dragonfly larva is able to shoot its 'mask' at small prey from exactly the distance at which it can actually seize it. When one of the compound eyes is blinded, the reaction can be elicited by objects of almost any size, provided the angle under which it is seen corresponds with the angle under which normal prey is seen at the right distance. In other words, distance reception has been lost (Baldus, 1926). The only species in which visual distance reception has been systematically analysed is man; here a number of other principles are involved besides the main one of binocular vision.

Distance discrimination with the aid of other sense organs is negligible. One of the few specialized cases is found in bats, where the ear is

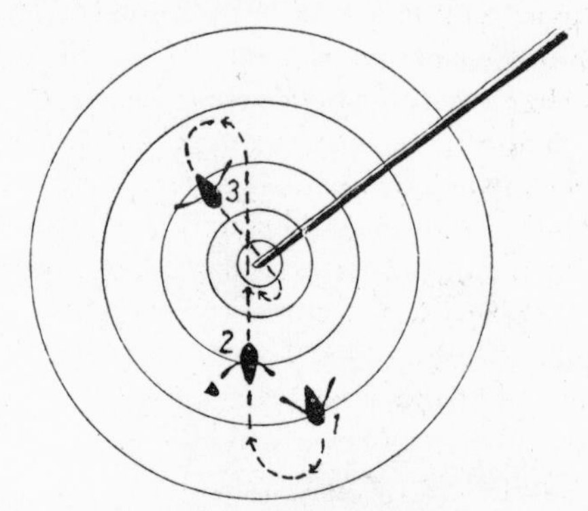

FIG. 14. *Notonecta glauca* reacting to surface ripples. After Baerends, 1939.

capable of judging the distance by means of 'echolocation' (Griffin; summary in Dijkgraaf, 1946).

The sense of touch, and sometimes touch and chemical receptors jointly, enable many animals to build up a three-dimensional world picture, but the working sphere never extends beyond that part of the surroundings which is in direct touch with the body.

Thus the marine fishes of the genus *Trigla* possess taste-buds on the three anterior rays of the pectorals. These rays are not connected by

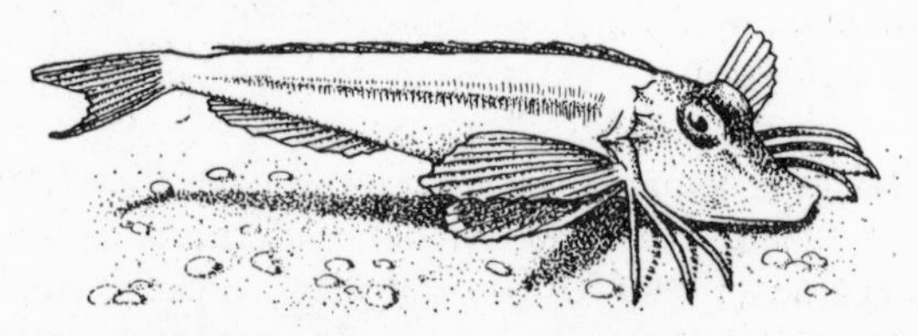

FIG. 15. *Trigla* 'walking' and testing substrate chemically.

skin with the other rays; they can be moved independently and thus enable the fish to localize food (Scharrer, 1935) (Fig. 15).

The lateral line organs found in many water-living vertebrates are an exception in that they permit localization of a distant object.

Many fish, for instance, have a complicated reflex system which enables them to locate and snap at a small prey even when they are blinded (Wunder, 1927). The fighting of the males of many species consists, in part, of directing, by means of the tail, a strong water current towards the opponent (Fig. 16). These tail blows, sometimes delivered from distances equalling the length of the fish, induce, under certain conditions, the opponent to flee. The very specialized lateral line organs of the aquatic toad *Xenopus laevis* enable it to locate a moving object up to a distance of 10 cm. (Kramer, 1933).

These two principles of spatial analysis of the environment, viz. localization of direction and of distance, are of great importance for the understanding of the influence of the environment on behaviour. First,

FIG. 16. Two fish fighting by aiming 'water jets' at each other's lateral line organs.

spatial analysis enables an animal to 'recognize' objects. Further, it enables it to localize objects in relation to other parts of the environment and thus to perform oriented movements, that is, movements directed in relation to spatial patterns outside the animal.

ACTUAL VERSUS POTENTIAL STIMULI

Sign Stimuli

A mere knowledge of the potential capacities of the sense organs never enables us to point out, in any concrete case, the actual complex of stimuli responsible for the release of a reaction. From a study of sensory capacity we can infer what changes in the environment can or can *not* be perceived by the animals, but a positive answer about what *does* release the observed reaction is impossible. This turns upon the peculiar fact that an animal does not react to all the changes in the environment which its sense organs can receive, but only to a small part of them. This is a basic property of instinctive behaviour, the importance of which cannot be stressed too much. For instance, the carnivorous water beetle *Dytiscus marginalis*, which has perfectly developed compound eyes (Fig. 17) and can be trained to respond to visual stimuli, does not react at all to visual stimuli when capturing prey, e.g. a tadpole. A moving prey in a glass tube never releases nor guides any reaction. The

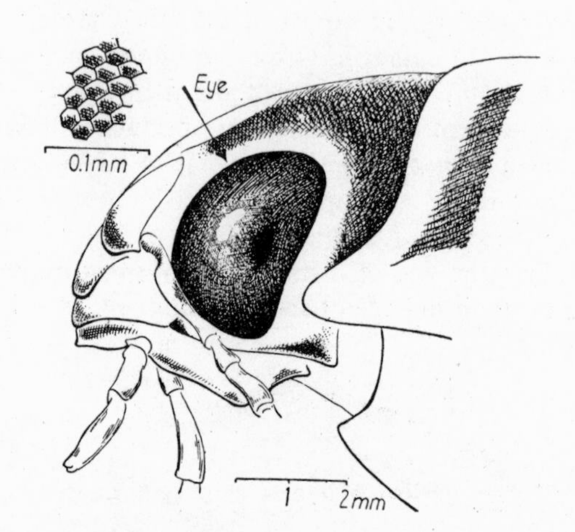

FIG. 17. Head of *Dytiscus marginalis* showing well-developed compound eyes. After Tinbergen, 1947.

FIG. 18. *Dytiscus* reacting to scent of prey and caught in a 'physiological trap'. After Tinbergen, 1936*c*.

beetle's feeding response is released by chemical and tactile stimuli exclusively (Tinbergen, 1936*c*); for instance, a watery meat extract promptly forces it to hunt and to capture every solid object it touches (Fig. 18).

The occurrence of such 'errors' or 'mistakes' is one of the most conspicuous characteristics of innate behaviour. It is caused by the fact that an animal responds 'blindly' to only part of the total environmental situation and neglects other parts, although its sense organs are

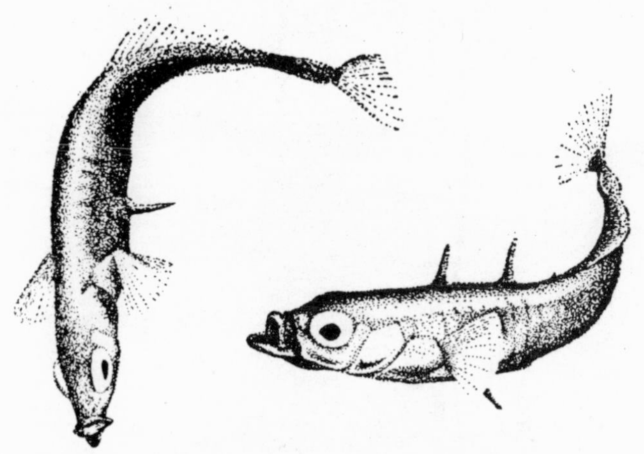

FIG. 19. Two male three-spined sticklebacks fighting. After Ter Pelkwijk and Tinbergen, 1937.

perfectly able to receive them (and probably do receive them), and although they may seem to be no less important, to the human observer, than the stimuli to which it does react. These effective stimuli can easily be found by testing the response to various situations differing in one or another of the possible stimuli. A small number of such experimental studies have been carried out; they have led to important results.

Moreover, even when a sense organ is involved in releasing a reaction, only part of the stimuli that it can receive are actually effective. As a rule, an instinctive reaction responds to only very few stimuli, and the greater part of the environment has little or no influence, even though the animal may have the sensory equipment for receiving numerous details.

For instance, the spring fighting of male sticklebacks (Fig. 19) is especially directed against other male sticklebacks in nuptial markings. As the males differ from other animals, especially in having an intensely red throat and belly, it seems probable that the red colour might be the most important stimulus. This has been tested in the following way.

Models of sticklebacks were presented to a number of males (Fig. 20). Some of the models were very crude imitations of sticklebacks, lacking many of the characteristics of the species or even of fish in general, but possessing a red belly (Series R). Others were accurate imitations of

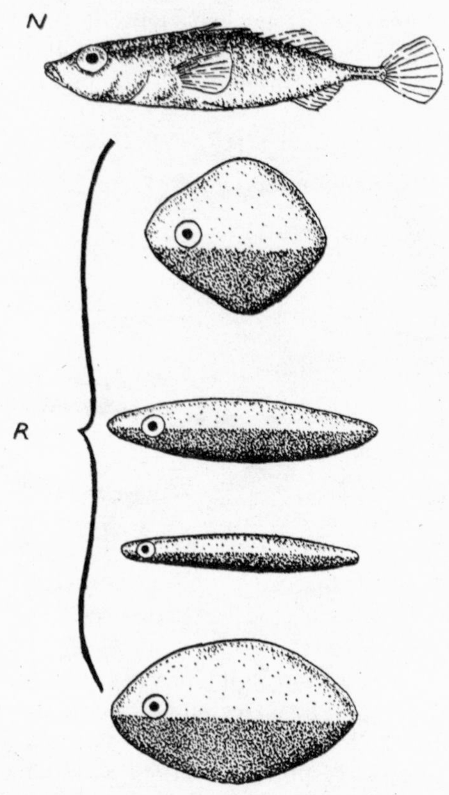

FIG. 20. A stickleback model of series *N* (above) and four of series *R*. After Tinbergen, 1948.

sticklebacks, but lacked the red (Series N). The males attacked the first group of models much more vigorously than they did the others. In this experiment the red colour was put into competition against all other morphological characters together. The results prove that the fish reacted essentially to the red and neglected the other characteristics. Yet its eyes are perfectly able to 'see' these other details (Ter Pelkwijk and Tinbergen, 1937).

Much the same condition exists in the English robin. Lack (1943) discovered that a territory-holding male of this species would threaten a mere bundle of red feathers much more readily than a complete mounted young robin which showed all the characteristics of a robin except the red breast. Again, the red breast is the effective stimulus (Fig. 21).

A somewhat more complicated case is the following. Newly hatched chicks of the herring gull beg for food by pecking at the tip of the parent's bill. The latter regurgitates the food on to the ground, picks up a small morsel and, keeping it between the tips of the beak, presents it to the young (Fig. 2, p. 7). After some incorrect aiming the young

FIG. 21. Two models of European robin. Right: a tuft of red feathers; left: mounted young robin with dull brown breast. After Lack, 1943.

gets hold of the food and swallows it. The bill of the herring gull is yellow, with a red spot at the end of the lower mandible. By comparing the chicks' reactions towards (1) a flat cardboard dummy in natural colours and (2) a similar dummy lacking the red patch, it was found that the red patch was of great importance. Further, a patch of any colour, black, blue, even white, gave the dummy a considerably higher releasing value than the dummy without any patch (Fig. 22). The fact that even a white patch increased the releasing value pointed to the conclusion that contrast between bill and patch played a part; the fact that red had more influence than even black indicated that red as colour had also influence. In order to test the first possibility, the series represented in Fig. 23 was presented. The bills were of a uniform grey of the same brightness in all the dummies; the patches varied from white to black in small steps. The results indicated in Fig. 23 show that contrast was part of the stimulus situation. A comparison of models with varying bill-colour (Fig. 24) shows that (*a*) red as such is important and (*b*) yellow has no influence at all. A last series, in which the colour pattern of the bill was constant but the colour of the head was varied, showed that a model with a white head had no more releasing value than models with black, red, yellow, green, blue, &c., heads. These observations lead to the conclusion that the chick reacts especially to the red patch. This

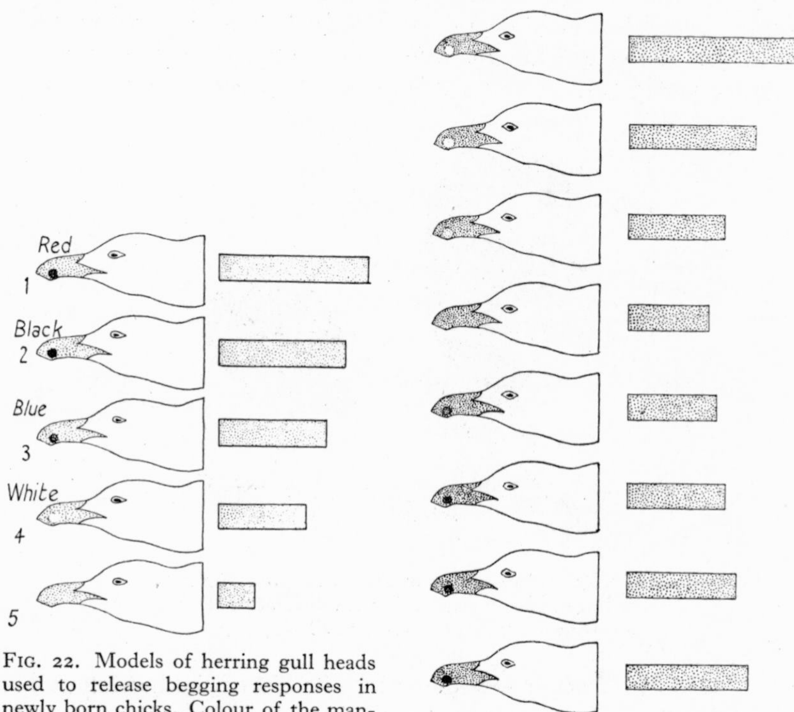

FIG. 22. Models of herring gull heads used to release begging responses in newly born chicks. Colour of the mandible patch varied (1–4) or absent (5). Columns indicate relative frequency of chicks' responses. After Tinbergen, 1949.

FIG. 23. Releasing value of herring gull models with grey bills with patches of varying shade. After Tinbergen, 1949.

patch works through its colour and through its contrast with the colour of the bill.

No releasing value was found for either colour of the bill or colour of the head (Tinbergen, 1948*b*, 1949; Tinbergen and Perdeck, 1950).

The reactions of many birds to flying birds of prey are often released by quite harmless birds. The domestic cock gives its alarm call, not only when a sparrow hawk is passing, but also as a reaction to the sudden appearance of a pigeon or a crow. The special type of movement, the sudden appearance, is sufficient to elicit the alarm, although the shape of a pigeon is quite different from that of any bird of prey. In addition, many birds react to the typical shape of a bird of prey in flight. Heinroth and Heinroth (1928) relate how many birds in the Berlin zoo react by escape to sailing swifts in the first days after the latter's arrival in spring. As the shape of a swift in flight is very similar to that of a bird of prey (Fig. 25)—both have a remarkably short neck—it seems that in this case the special shape accounted for the erroneous reaction. In order to test this hypothesis, several workers have studied the reactions of birds to cardboard models of flying birds (Goethe, 1937; Krätzig, 1940;

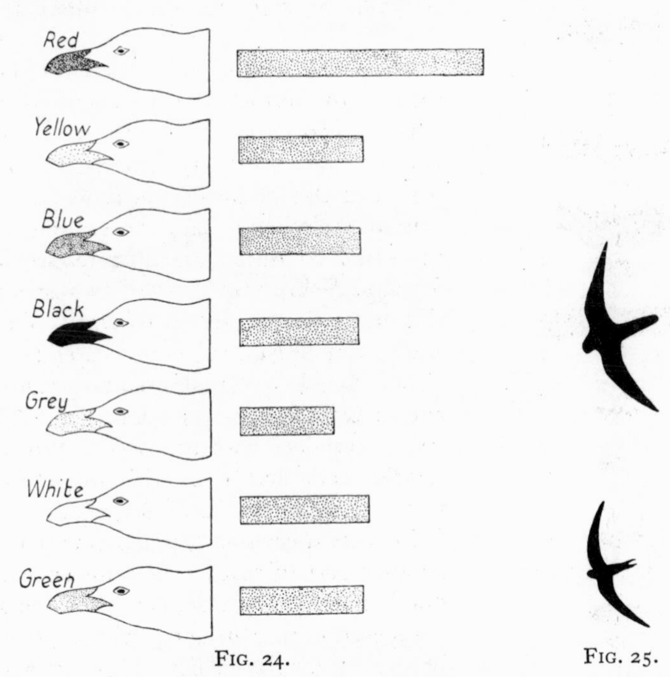

FIG. 24. FIG. 25.

FIG. 24. Releasing value of herring gull models with uniform bills of varying colour. After Tinbergen, 1949.

FIG. 25. Flying hobby (*Falco subbuteo*) (above) and swift (*Apus apus*) (below).

Lorenz, 1939). These tests all showed that, as long as a model had a short neck, the experimental animals (various species of gallinaceous birds, ducks, and geese) would show alarm. Other characteristics, e.g. shape and size of wings and tail, were rather irrelevant (Fig. 26). This indicates, therefore, that the errors described are due to the birds' reacting to only one out of a number of possible stimuli.

In the visual domain, motion may often be a powerful stimulus. One of the earliest studies of this type concerned the 'recognition' of prey by dragonflies (Tirala, 1923). According to this author, mosquito-hunting species do not react to properties of shape, although their highly developed compound eyes certainly enable them to see even minor differences in shape. They react specially to the type of motion of flying mosquitoes. Mosquitoes are not hunted when walking on solid ground. Small scraps of paper of varying shape but of approximately the right size promptly release the hunting responses when they are thrown in the air.

These examples concern visual stimuli. Numerous instances are known of the restriction of sign stimuli to other sensory fields.

Striking examples of restriction to chemical stimuli are found in the reactions of the males of certain Noctuid moths to the sexual odours emanated by the females. In *Saturnia pyri*, and also in *Lymantria dispar*, in *Lasiocampa* species, and many other species, males in sexual condition are attracted by virgin females. Fabre was the first to suspect that this must be a reaction to smell. This has since been proven in several cases (see Von Frisch, 1926). The males react so vigorously and so exclusively to the odour that they may try to copulate with any object bearing the female scent and even with the object on which a female has just been sitting.

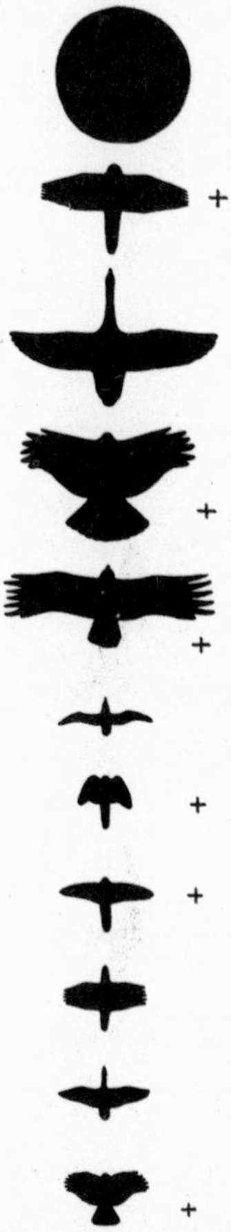

FIG. 26. Bird models used by Lorenz and Tinbergen for testing reactions of various birds to birds of prey. Those marked + released escape responses. After Tinbergen, 1948.

In other species of Lepidoptera scent plays another part in mating. For instance, in the grayling the male stimulates the female to co-operation in mating by bringing the scent organs of his forewings (Figs. 27 and 28) in touch with the female's chemoreceptors, which are located on the antennae. This display takes place after the sexual pursuit (p. 40), when the female has alighted and the male has taken up a position in front of her. The climax of his elaborate courtship is an elegant bow (Fig. 29) by which the female's antennae are caught between the male's forewings. Males in which the scent organs have been removed have great difficulty in acquiring a mate in spite of intensive courting (Tinbergen, Meeuse, Boerema, and Varossieau, 1942).

Reactions to sound may also be strikingly independent of other possible stimuli. Brückner (1933) studied the social relationships in domestic fowl. He found that a hen coming to the rescue of a chick in

Fig. 27. The grayling. Upper: dorsal side; lower: ventral side. Black line in upper left figure indicates position of scent organ on left wing. After Tinbergen, Meeuse, Boerema, and Varossieau, 1942.

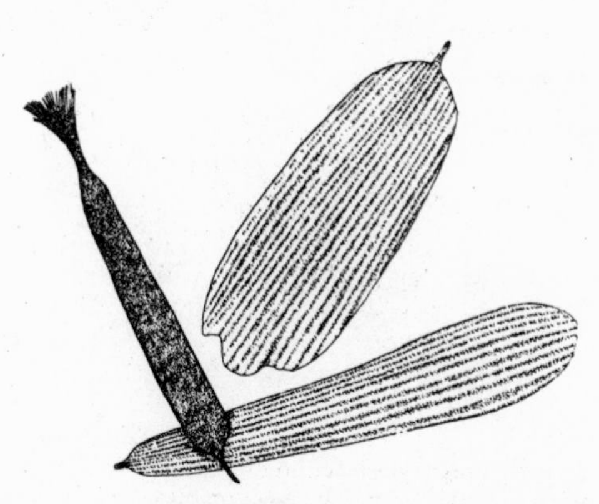

Fig. 28. Two ordinary wing scales and one 'scent scale' of male grayling. After Tinbergen, Meeuse, Boerema, and Varossieau, 1942.

distress is reacting to the distress call, not to the chick's movements. When he fastened a chick to a peg, keeping it out of sight by putting it behind a screen, the mother would come to its rescue when she heard

FIG. 29. Grayling male (right) bowing so that the female's antennae come in contact with the scent organ of the male. After Tinbergen, Meeuse, Boerema, and Varossieau, 1942.

FIG. 30. Brückner's experiments on reaction of domestic fowl to distress call of chick. Above: visual stimulus presented alone; no reaction observed. Below: auditory stimulus presented alone; intensive reaction.

the chick whining. But when the chick was put under a glass dome in full view, so that the mother could see it struggling but could not hear its distress notes, she was entirely indifferent (Fig. 30).

In locusts of the species *Ephippiger ephippiger*, females that are willing to mate wander to the singing males. Whereas they are attracted to invisible singing males from at least 10 yards distance, they ignore silent males even when quite near. Males in sexual condition that were silenced

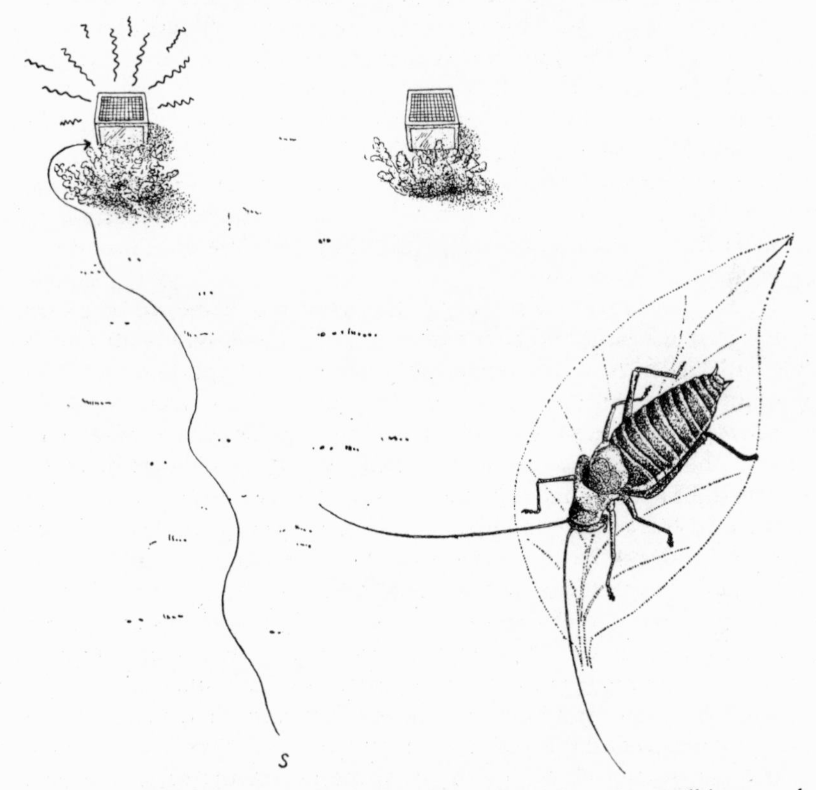

FIG. 31. *Ephippiger ephippiger*. Male (right). Path of female in mating condition towards cage with singing males (left), neglecting cage with silenced males. After Duym and Van Oyen, 1948.

by gluing their wings together, a minor operation, were not able to attract a single female (Duym and Van Oyen, 1948) (Fig. 31).

Touch receptors may also have very specific releasing functions. Fighting in male sticklebacks may consist of their repeatedly biting each other. This response is released by one hitting the other with its snout. It can easily be evoked by imitating this tactile stimulus with a glass rod or any other solid object. Whereas fighting as a whole is dependent on visual stimulation by a male in nuptial markings, the release of this specialized part of the fighting pattern is almost or perhaps entirely independent of visual stimuli.

It is not necessary to carry this review on; facts of a similar nature will be described in several of the subsequent chapters of this book.

Russell (1943), who has published a valuable review of these and similar facts in many different animals, has called these essential stimuli 'sign stimuli'. Later he preferred 'perceptual signs'. For reasons not to be discussed here, I shall use the term 'sign stimuli', although I am quite aware that the term 'stimulus' is open to certain criticisms. As a provisional, descriptive term, however, it will do.

As far as the available facts go, this dependence on only one or a few sign stimuli seems to be characteristic of innate responses.

In every study of reactive behaviour it is essential to be well aware of this difference between what an animal can perceive and what it actually reacts to in a given case. Neglecting this difference may lead to gross misrepresentation.

Thus Allen (1934), in a study of the courtship of the ruffed grouse, found that males in sexual condition copulated not only with females but with males as well, provided they assumed a position more or less resembling the female's normal mating position. Thus 'a stuffed grouse, a grouse skin or a dead grouse' released the copulatory response in any male. 'The exact pose was unimportant so long as it was more or less flattened, or at least not mounted in an attitude of display, and the sex of the bird was equally unimportant' (Allen, 1934, p. 192). From these facts Allen drew the conclusion that the ruffed grouse male does not distinguish between sexes. The facts, however, merely show that the copulatory response of the male is released by a stimulus situation in which no morphological sign stimuli play a part. The crouched position of the willing female is the most important sign stimulus. Allen's conclusion, therefore, is too general in two respects: first, males do distinguish between the sexes, but in this reaction they use behaviour characters instead of differences of shape or colour; second, even if they do not react to morphological properties, they may be quite able to see them and hence to distinguish between the sexes. A crouching submissive male or a dead male release the copulatory response merely because the male cannot resist the powerful sign stimulus.

In a similar way, the reactions of insects to the colours of flowers have been misinterpreted. Hive bees (von Frisch, 1914), some flies (*Bombylius*, Knoll, 1921–6), butterflies (Ilse, 1929; Tinbergen *et al.*, 1942), hawk moths (Knoll, 1921–6), and other insects specialized in sucking nectar have been shown to react innately to blue and yellow objects. Although they may react selectively to either blue or yellow, they do not show preference for special hues within the blue-violet-purple group nor within the wide range of orange-yellow-yellowish-green. It has been inferred that they cannot discriminate between colours within each of these groups, but Lotmar (1933) showed convincingly that hive bees readily distinguished between different hues if specially trained to show differential responses. The recognition of the peculiar nature of sign stimuli as being different from the potential stimuli that can be received by the sense organs prevents confusion of this sort.

The reason why the dependence of innate behaviour on sign stimuli has not yet been generally recognized probably lies in the fact that so many laboratory psychologists have been studying conditioned reactions. Conditioned reactions are, so far as we know, not usually dependent on a limited set of sign stimuli, but on much more complex stimulus situations. I shall return to this problem in Chapter VI.

It is the dependence of innate behaviour on sign stimuli that renders it possible to evoke reactions in an animal by presenting it with dummies. As a matter of fact, when any animal readily responds to a dummy, this is a certain indication that its reaction is dependent on sign stimuli.

As mentioned above, the distinction between *Umwelt*, especially *Merkwelt*, and environment (von Uexküll) was partly based on the fact that different species have different sensory capacities. In this paragraph we have found a further justification of this distinction. The animal's own world is not only dependent on what its sense organs can or cannot receive. Its sensory world is still more restricted; it is composed of sign stimuli, at least as long as we are dealing with innate responses. This implies that the animal's perceptual world is constantly changing and depends on the particular instinctive activity that is brought into play.

The 'Innate Releasing Mechanism'

Up to this point I have been purposely simplifying matters by confining myself to pointing out that the animal does not respond to many characteristics of a situation, and that there are but few essential sign stimuli. I did not try to find out whether the sign stimuli I mentioned were the only effective stimuli. Our next task will now be to find the optimal stimulus situation of a given reaction. Some reactions of the three-spined stickleback may again be taken as examples.

As I showed, the males' fighting response is dependent on the sign-stimulus 'red belly'. The dummy tests described, however, were not adapted to study the influence of the movement, because, in every test, the dummies were either all moved in the same way or kept motionless. Now a male stickleback often moves in a very special way. When encountering another male near the boundary of its territory (which is where most of the fighting takes place) it adopts a posture with the head pointed downward, holding itself in a very peculiar vertical position. Now it can easily be shown that a dummy will evoke a much more vigorous attack when it is presented in this 'threatening' position than when shown in a normal position (Fig. 32). The fighting response of a male stickleback, therefore, is not only released by a red male, but also by the special movements (posturing) of a male. The response, therefore, is dependent on a combination of these two sign stimuli.

The courting behaviour of a male stickleback before a pregnant

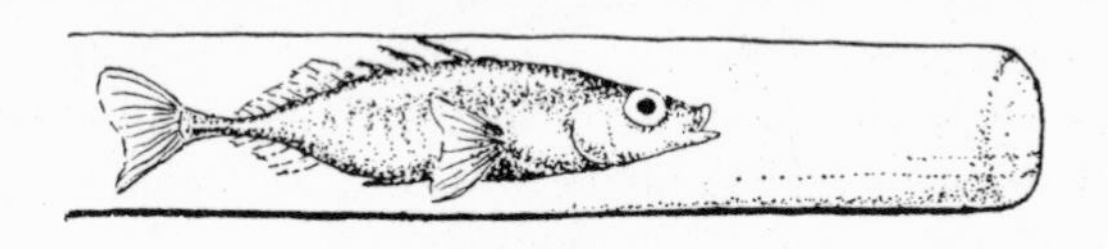

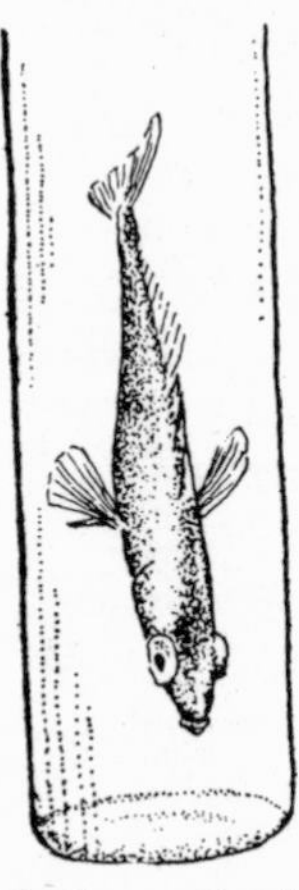

Fig. 32. Male three-spined stickleback prevented from assuming threat posture (above) and threatening (below). After Ter Pelkwijk and Tinbergen, 1937.

female is also dependent on at least two sign stimuli: the swollen abdomen and the special posturing movement of the female. When a crude fish-like model with a swollen abdomen is presented to the male, it will vigorously court this ridiculous dummy, whereas its response to a complete stickleback which has a normal belly is much less intense (Fig. 33). Also, a dummy that is posturing after the manner of a female (Fig. 34) releases the male's courtship much more readily than when it is presented in normal position.

The female's reaction to the courting male is released by two sign stimuli: the red belly and the male's special movements, the 'zigzag dance'.

It is necessary to insert here a few remarks about the technique of these dummy experiments. As will be discussed more fully in a later chapter, these reactions are not controlled by external stimuli exclusively, but also by the internal reproductive drive. In the autumn or in winter the best dummies will invariably fail to evoke these responses. This is because the drive, or motivation, is too low in intensity. If the drive is of medium intensity a relatively strong stimulus situation will be needed to get a response at all; if the drive is very strong, the slightest stimulation will be followed by an explosive reaction. Under such condi-

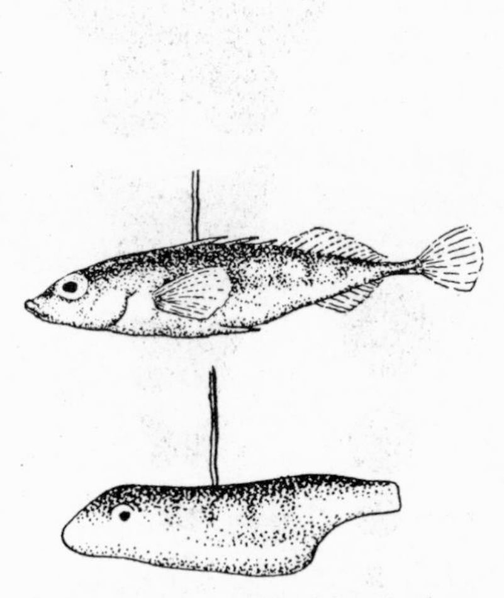

FIG. 33. Two models of female three-spined stickleback. Detailed model with 'neutral' abdomen (above); crude model with swollen abdomen (below). After Tinbergen, 1942.

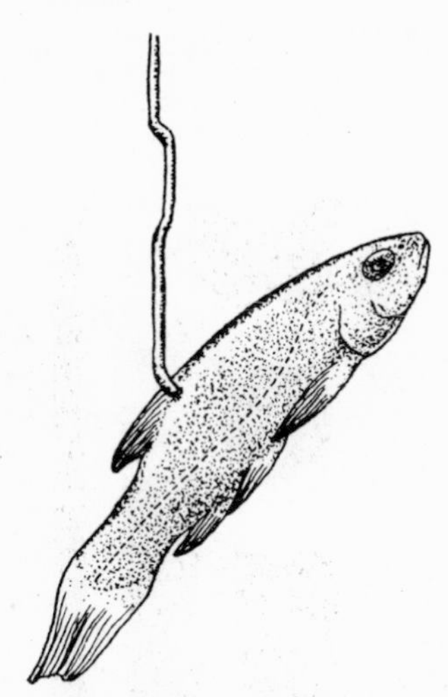

FIG. 34. Dead tench (*Tinca vulgaris*) of stickleback size presented in attitude of readiness of female three-spined stickleback. After Ter Pelkwijk and Tinbergen, 1937.

tions a female will respond to a dummy that displays only one of the sign stimuli, e.g. the zigzag dance and not the red colour. This explains, for instance, why Leiner (1929, 1930) could get his animals to spawn in monochromatic light of various colours. However, to infer, as Leiner did, that the red colour has no influence in releasing the female's reaction, is a mistake. The monochromatic light test merely shows that red is not altogether indispensable for females with an exceptionally strong drive.

In an experiment of this kind, as in every experiment, it is necessary to compare the reactions to two different situations with each other. These two situations must differ only in the one factor the influence of which is to be studied. In our case we have to compare a dummy displaying the sign stimuli *A* and *B* with a dummy showing only one of them, viz. either *A* or *B*. Now such a test may have results differing with the intensity of the drive. If the drive is weak, both models will fail to evoke a response. If the drive is strong, both models will give a response; and although the response to $A+B$ may differ in intensity from that to A, the difference is often difficult to detect, because it is only a difference of degree. But when the drive is of medium intensity the animal will show a positive response to $A+B$ and no response at all to A. It is the experimenter's job so to choose his conditions and his animals as to get

FIG. 35. Grayling models of varying shape. After Tinbergen, Meeuse, Boerema, and Varossieau, 1942.

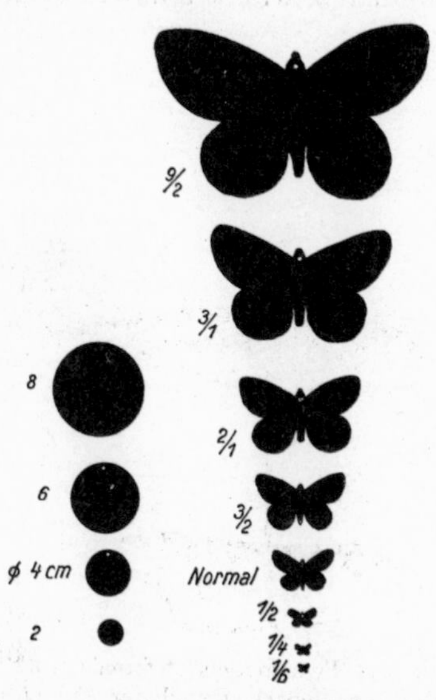

FIG. 36. Grayling models of varying size. After Tinbergen, Meeuse, Boerema, and Varossieau, 1942.

this difference out as clearly as possible. Of course this is not cheating; the difference can be seen under the other conditions, but it is merely less easy to see and above all to describe.

More or less complete studies of all the sign stimuli that affect one single reaction have been carried out in a few cases. One is the mating flight of the male of the grayling (*Eumenis semele*), a satyrid butterfly. The male takes the initiative in mating by pursuing a passing female in flight. A virgin female thus approached alights, and the male performs an elaborate series of instinctive 'ceremonies' which eventually lead to mating. The first reaction, the sexual pursuit, has been studied by means of dummies, in which shape (Fig. 35), size (Fig. 36), colour, light intensity, type of movement (Fig. 37), and distance were varied. The result was that neither colour nor size or shape were of much influence, but that light intensity (darkness), type of movement, and distance had a profound influence, the 'optimal' female being an object that flies in the typical fluttering way of a butterfly, that is as dark as possible and as near as possible (Tinbergen, Meeuse, Boerema, and Varossieau, 1942).

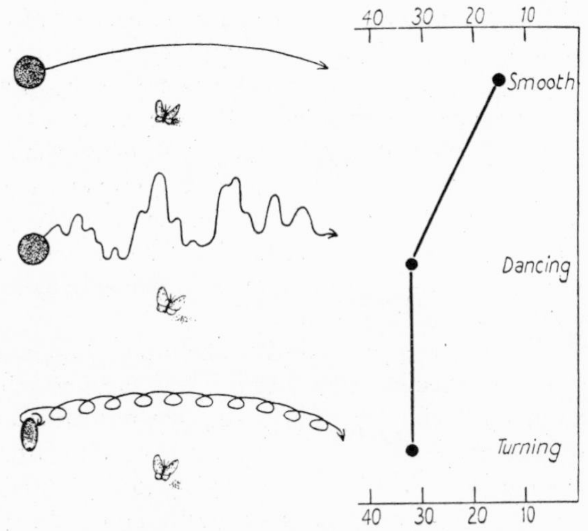

Fig. 37. Influence of type of motion on sexual pursuit of male grayling. Adapted from Tinbergen, Meeuse, Boerema, and Varossieau, 1942.

Fig. 38. Gaping reaction of young thrushes. After Tinbergen, 1947*b*.

The sign stimuli releasing the gaping reaction of young thrushes of about 10 days of age (Fig. 38) are the following: the object (the parent bird) has to move, it may have any size above about 3 mm. in diameter, and it must be above the horizontal plane passing through the nestlings' eyes. Optimal dummies presented below that plane may be seen, as can be judged from eye movements, but they never release the gaping reaction (Tinbergen and Kuenen, 1938).

The strict dependence of an innate reaction on a certain set of sign stimuli leads to the conclusion that there must be a special neuro-sensory mechanism that releases the reaction and is responsible for its selective susceptibility to such a very special combination of sign stimuli. This mechanism we will call the Innate Releasing Mechanism (IRM), a free translation of the German term *das angeborene auslösende Schema* (Von Uexküll–Lorenz).

As we said before, the fact that so many animals do react to only a few sign stimuli at any one time has the practical implication that, if we know the potential capacities of the sense organs of a given species, this certainly does not mean that we know the external causes of any particular reaction. It does imply, in the second place, that an animal's failure to respond to certain changes in the environment does not prove inability to perceive those changes under any circumstances. It merely proves that they do not influence the IRM of the reaction studied. This is one of the reasons why conditioning is so valuable as a method of studying the sense-organs' capacities.

There is some evidence which tends to show that there is no absolute distinction between effective sign stimuli and the non-effective properties of an object. Lack (1943) found that the posturing of the robin is not exclusively dependent on the sign-stimulus 'red breast', for, rarely, a specimen lacking red on the breast was postured at. Similarly, I found in the three-spined stickleback that dummies lacking red and presented in neutral position would sometimes release attack, though a feeble one. These observations suggest that dependence on a sharply limited number of sign stimuli might represent an extreme case and is, perhaps, a specialization. Another possibility is that conditioning is responsible for the effectiveness of additional stimuli. More experimental studies of IRMs are necessary to elucidate this problem.

Although up till now few IRMs have been studied adequately, what scanty knowledge we have is sufficient to show that in general no two reactions of a species have the same IRM.

As already mentioned, the mating pursuit of the male grayling is released by a stimulus situation in which colour takes no part. The natural conclusion to be drawn from this would seem to be that *Eumenis* is colour-blind. But the observation that *Eumenis* selects blue and yellow flowers to feed on seems to contradict this. When the feeding reactions were analysed in the ordinary way, by presenting the butterflies with paper flowers of standardized coloured and grey papers, *Eumenis* appeared to be able to react quite well to yellow and blue on the basis of a real colour-discrimination (Fig. 39). Here then was a clear-cut case showing that an animal may react to colours in one reaction while not 'distinguishing' between them in another reaction. Of course, this very same state of affairs was the cause of the dispute between von Hess and von Frisch regarding the reactions of the honey bee to colours, which was described above.

Similar results have been obtained with two other species of Lepidoptera. Knoll (1921, 1926) found that the hawk moth *Macroglossa stellatarum* selects yellow and blue objects when hungry, yellowish-green objects when selecting a place to deposit eggs, and dark objects of any colour or grey when selecting a crevice for the purpose of hibernating.

Pieris brassicae selects yellow, blue, and red flowers for feeding, but for oviposition the female selects green objects (Ilse, 1929).

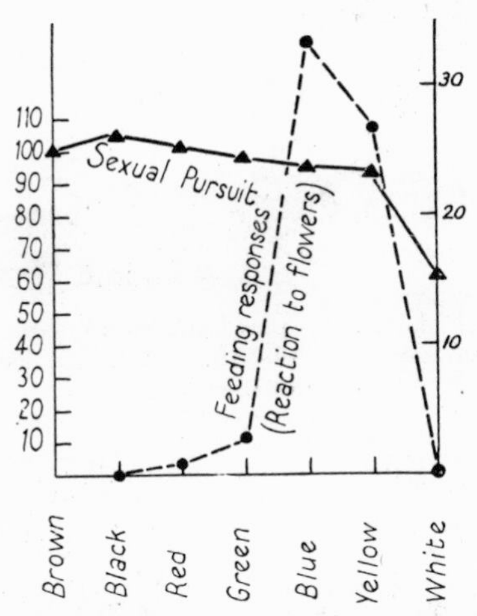

FIG. 39. Reaction of grayling to colour in two different motivational conditions. After Tinbergen, Meeuse, Boerema, and Varossieau, 1942.

These few instances will suffice to illustrate the general conclusion, viz. that different reactions of the same animal have different releasing mechanisms. This conclusion is still more obvious when one studies different reactions of an animal to the same object. When the female stickleback reacts to a courting male by posturing to him, she responds to his red colour and the special movement of the zigzag dance. But when, within one or two seconds, she enters the nest, her spawning reaction, although equally dependent on stimulation by the male, is released by quite different stimuli. As soon as she enters the nest, the male begins to thrust its snout at her rump with quick, rhythmic movements (Fig. 40). When the male is taken away, the female is absolutely incapable of spawning. But when the experimenter then substitutes a glass rod or any hard object for the male and gives her the same mechanical stimulus, she will respond by spawning. Thus the same object (the male) has to provide the female with entirely different stimuli for the two reactions.

It is not necessary to elaborate this point further. A great number of facts of this kind are given by Russell (1943).

'Supernormal' sign stimuli

The innate releasing mechanism usually seems to correspond more or less with the properties of the environmental object or situation at

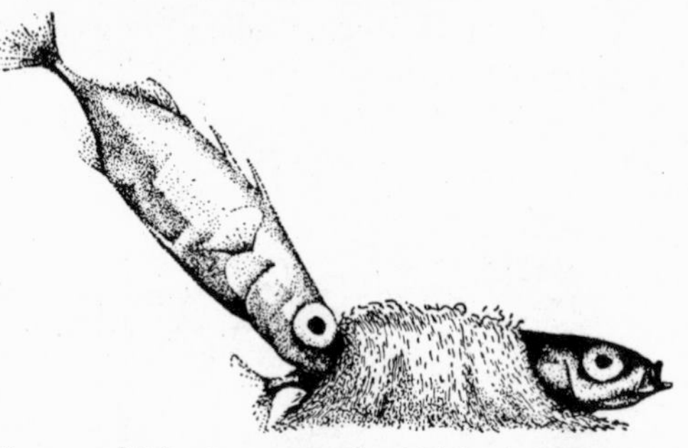

FIG. 40. Male three-spined stickleback stimulating female to spawn by 'quivering'. After Ter Pelkwijk and Tinbergen, 1937.

FIG. 41. 'Supernormal' egg (left) preferred by ringed plover (*Charadrius hiaticula*) to normal egg (right). Courtesy of Prof. O. Koehler.

which the reaction is aimed. This is according to expectation, and would even seem to be a truism, for otherwise the reaction would run the risk of being released by the 'wrong' situation and chaos would result.

However, close study of IRMs reveals the remarkable fact that it is sometimes possible to offer stimulus situations that are even more effective than the natural situation. In other words, the natural situation is not always optimal.

This was first discovered by Koehler and Zagarus (1937) in a study of 'egg recognition' (or the external stimuli releasing reactions normally released by the eggs) in the ringed plover. If presented with a normal egg (which is light brownish with darker brown spots) and an egg with a clear white ground and black dots (Fig. 41) the birds preferred the latter type.

In a similar way we found that oystercatchers preferred a clutch of five eggs to the normal clutch of three (Fig. 42). Still more astonishing is the oystercatcher's preference for abnormally large eggs. If presented

FIG. 42. Oystercatcher (*Haematopus ostralegus*) incubating 'supernormal' clutch of five eggs in preference to natural clutch of three.

FIG. 43. Oystercatcher reacting to giant egg in preference to normal egg (foreground) and herring gull's egg (left). After a photograph in Tinbergen, 1948.

with an egg of normal oystercatcher size, one of herring gull's size, and one double the (linear) size of a herring gull's egg, the majority of choices fall upon the largest egg (Fig. 43).

Another instance is the male grayling's sexual pursuit flight. As was related above, dummies of females of different colours had about the same releasing value. There is, however, a slight difference. The darker

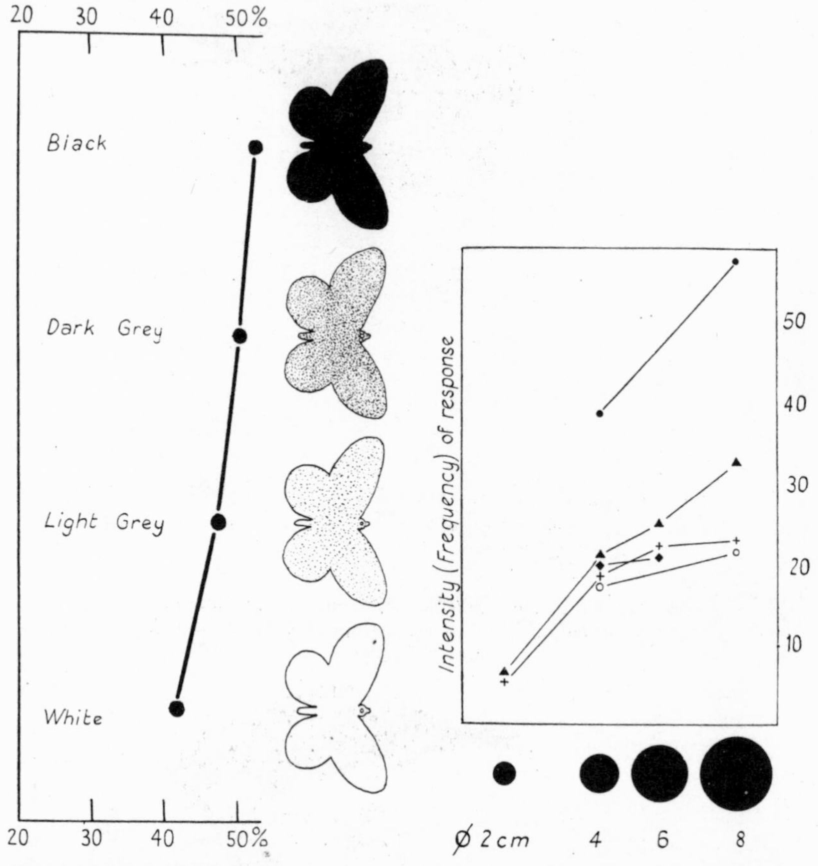

FIG. 44. Effectiveness of grayling models of various shades of grey. Changed after Tinbergen, Meeuse, Boerema, and Varossieau, 1942.

FIG. 45. Releasing value of circular models of varying size as presented to male grayling. After Tinbergen, Meeuse, Boerema, and Varossieau, 1942.

colours get more responses than the lighter ones. If models of different shades, varying from white to black, are presented, the darker shades get progressively more responses. A black model even evokes more reactions than a model in natural colours (Figs. 39, 44). Moreover, models of much greater size than normal get more responses than models of normal size (Fig. 45).

The full significance of the phenomenon of 'supernormal' sign stimuli is not yet clear. A closer study might well be worth while.

Reaction chains

Consistent study of the dependence of behaviour on sensory stimuli has further revealed the fact that many reactions, even relatively short and simple ones, are in reality a chain of separate reactions each of which is dependent on a special set of sign stimuli. Usually the first indication one gets of the chain character of a response is a sudden break during its progress. Such an abrupt break can be prevented by presenting a new stimulus situation of the special kind required at the right instant. The reactions of foraging honey bees to flowers, for instance, begin with a response to a visual stimulus in which colour plays an important part. Yellow and blue paper models of flowers especially attract bees from a considerable distance. However, a bee rarely alights on these models; at a distance of about 1 cm. it will hesitate and then lose interest. But if an odour of the right kind is added to the model, the next link in the chain is released: the bee settles down on the model and searches for nectar. In the complete response, this second reaction has to be followed by a third reaction, the insertion of the mouth parts into the flower and the consequent reaction of actually sucking nectar. These reactions depend on visual, tactile, and chemical stimuli, the exact part played by each of which has not been studied. What fragmentary information we have, however, shows that we have to do with a relatively long chain of reactions (von Frisch, 1927).

The hunting behaviour of the bee-hunting digger wasp *Philanthus triangulum* gives us another example (Fig. 46). A hunting female of this species flies from flower to flower in search of a bee. In this phase she is entirely indifferent to the scent of bees: a concealed bee, or even a score of them put out of sight into an open tube so that the odour escaping from it is clearly discernible even for the human nose, fails to attract her attention. Any visual stimulus supplied by a moving object of approximately the right size, whether it be a small fly, a large bumble bee, or a honey bee, releases the first reaction. The wasp at once turns her head to the quarry and takes a position at about 10–15 cm. to leeward of it, hovering in the air like a syrphid fly. Experiments with dummies show that from now on the wasp is very susceptible to bee-scent. Dummies that do not have bee-odour are at once abandoned, but those dummies that have the right scent release the second reaction of the chain. This second reaction is a flash-like leap to seize the bee. The third reaction, the actual delivery of the sting, cannot be released by these simple dummies and is dependent on new stimuli, probably of a tactile nature (Tinbergen, 1935).

One of the most complete analyses of chain reactions of this type has been carried out with the mating behaviour of the three-spined stickleback (Fig. 47). Fig. 48 summarizes the results. Each reaction of either male or female is released by the preceding reaction of the partner. Each arrow represents a causal relation that by means of dummy tests has actually been proved to exist. The male's first reaction, the zigzag dance, is dependent on a visual stimulus from the female, in which, as already mentioned, the sign stimuli 'swollen abdomen' and the special

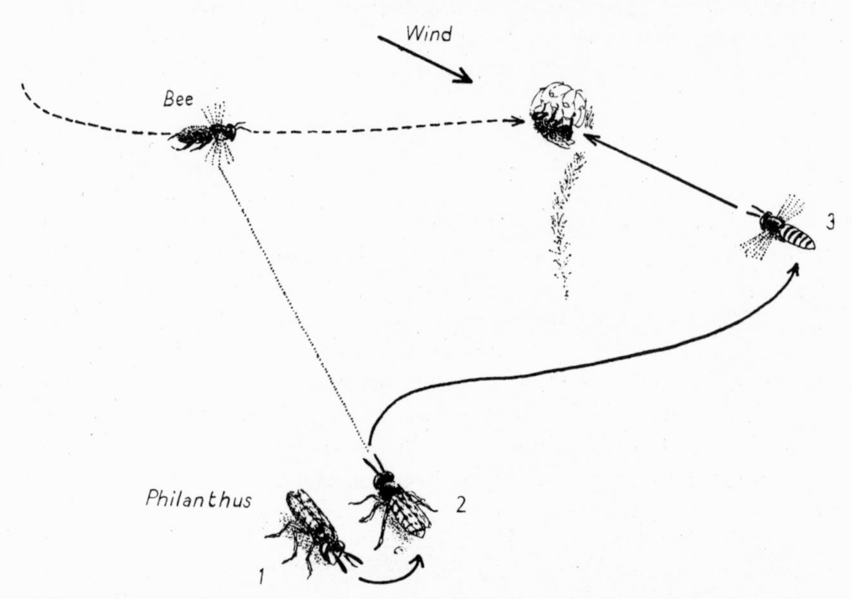

FIG. 46. Sequence of hunting behaviour of *Philanthus triangulum*. After Tinbergen, 1935.

movement play a part. The female reacts to the red colour of the male and to his zigzag dance by swimming right towards him. This movement induces the male to turn round and to swim rapidly to the nest. This, in turn, entices the female to follow him, thereby stimulating the male to point its head into the entrance. His behaviour now releases the female's next reaction: she enters the nest. As described above, this again releases the quivering reaction in the male which induces spawning. The presence of fresh eggs in the nest makes the male fertilize them.

Most of the links in these two reaction chains are dependent on visual sign stimuli, which are different for each of the links. The spawning depends on tactile stimuli. The male's ejaculation of sperm depends on a situation in which chemical and presumably tactile stimuli play a part (Ter Pelkwijk and Tinbergen, 1937; Tinbergen, 1942).

No doubt much the same state of affairs exists in most mating

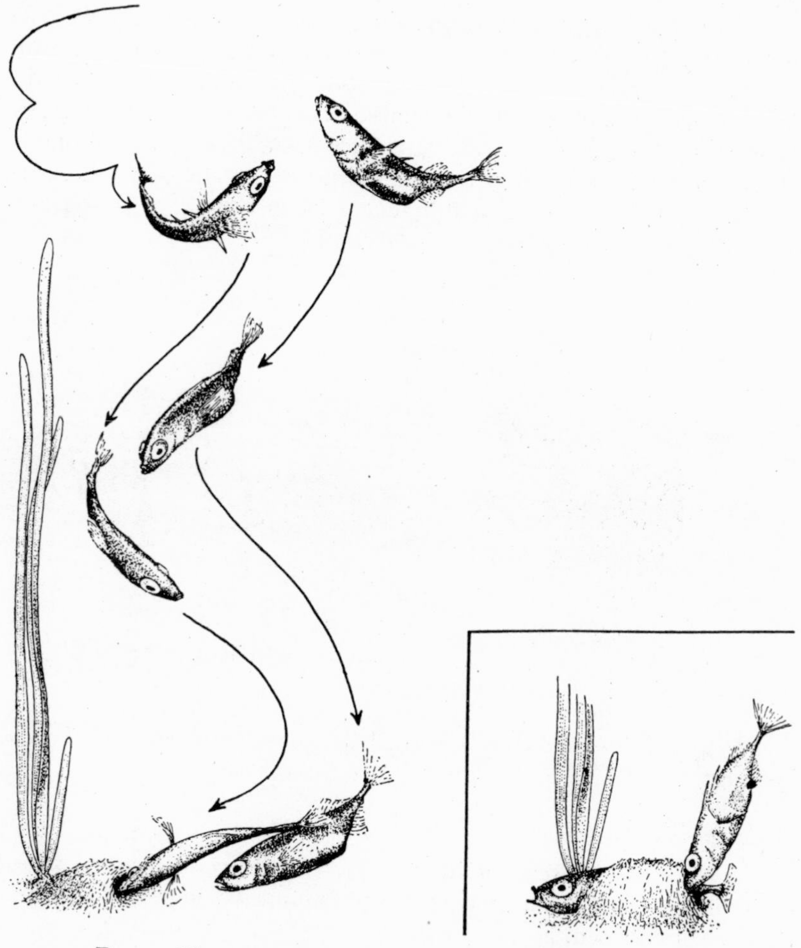

FIG. 47. The mating behaviour of the three-spined stickleback.

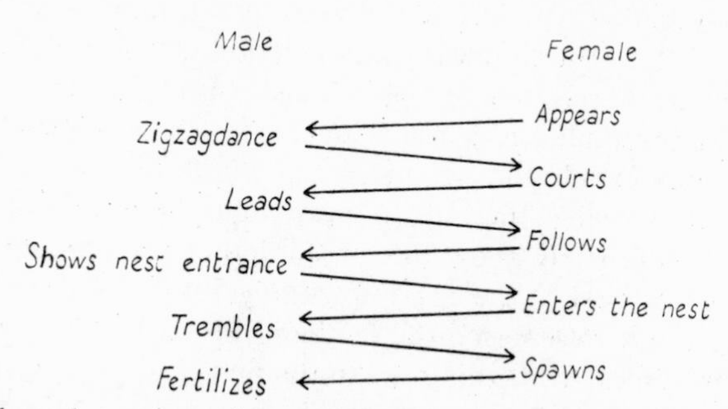

FIG. 48. Schematic representation of the relations between male and female three-spined stickleback. After Tinbergen, 1942.

behaviour. The long and complicated mating behaviour of the snail *Helix pomatia* L. (Fig. 49), reaching a climax in the release of the love-shaft by which the partners mutually stimulate each other to perform the final act, has been shown by Szymanski (1913) to be a reaction chain. By imitating the tactile stimuli delivered by the movements of one of the partners he could get the other partner to go through the entire series of reactions. The case is different from that of the stickleback in

FIG. 49. Mating *Helix pomatia* (left); love-shaft enlarged (right). After Meisenheimer, 1921.

that *Helix* is hermaphrodite, and the behaviour of the two partners is identical.

Ambivalent Behaviour

The fact that each reaction has its own releasing mechanism may lead to ambivalent behaviour when two sign stimuli belonging to different reactions are present at the same time.

In the breeding season a herring gull reacts to every red object in the nest by carrying it away, a reaction noticed by several observers, though its function is not entirely clear. The reaction of sitting down on the nest to incubate is released by sign stimuli from the eggs. The most important sign stimulus is a visual one. The shape is essential: any object of approximately the size of an egg and having a rounded form is accepted. Now when the gull is given a bright red dummy of egg shape, it will alternately show the two different reactions in incipient form: first it may peck at the egg and try to get it out of the nest, in the next instant it may raise its ventral feathers and settle down on the egg. The two sign stimuli 'something red in the nest' and 'egg-shaped object in the nest', respectively, were, as it were, struggling for priority, each activating a different action.

Innate Behaviour in Mammals

The work done with mammals, especially the rat, has been reviewed by Lashley (1938). His conclusions, though agreeing with the views set forth in this book in most essentials, differ in one important aspect. Contrary to our conclusion that an instinctive reaction is dependent on a limited number of sign stimuli, Lashley concludes that 'the accumulated observations suggest that the instinctive behaviour is dependent upon a complex of stimuli' (1938, p. 454), and that 'the stimulus is not a characteristic colour or odour, but seems to be a pattern, having the same characteristics of organization which we have found in studies of visual discrimination of objects' (p. 455).

In my opinion the available facts do not yet allow us to see whether this difference of opinion is due to a real difference in the behaviour; it seems quite probable that mammals are different from lower vertebrates and invertebrates. However, part of the discrepancy is certainly due to the fact that in the work reviewed by Lashley the chain character of the reactions was not sufficiently realized and investigated. If, for instance, we should study the mating behaviour of the male stickleback as a whole, we should find that many properties of shape, motion, colour, and even tactile and chemical properties of the female play a part in releasing the whole reaction. Analysis of the reaction, however, showed that mating, in this case, is a chain of separate reactions each of which is dependent on only a minor part of this complex of stimuli. Thus Lashley's conclusion is absolutely valid for the mating response as a whole, but it may be untrue of each of the separate elements of the chain.

It is certainly necessary to attempt a much more detailed analysis of 'mating behaviour', 'maternal behaviour', &c., in the rat before it will be possible to draw a conclusion about the nature of external releasing stimuli.

In other respects, too, students of innate behaviour have so far given insufficient attention to mammals. Yet the study of mammalian behaviour, differing as it does from the simpler type found in, for example, birds and fish, would be a test of the general applicability of our conclusions. Of course, it is also indispensable for a better understanding of human behaviour. The highly interesting study by Schenkel (1947) on the behaviour of the wolf demonstrates the possibilities of the ethological approach.

Innate or conditioned?

So far no mention has been made of the problem of determining which of the reactions discussed were innate and which were not. It is often impossible to judge this from observation of the adult animal. For instance, in all the species where the parents take care of the young, the behaviour of the latter may be conditioned by the adults in a number of ways. But an individual may also learn from experiences with other parts of the environment, such as food or predators. As we shall see in Chapter VI, learnt behaviour is by no means rare in the majority of species.

The only way to find out what behaviour is innate and what is acquired during individual life is to raise individuals in isolation, to observe the development of their behaviour, and to study the influence of different environments upon it. Various aspects of learning by individual experience will be discussed in Chapter VI. Here, however, something must be said about the criteria that allow us to recognize innate behaviour.

First, we must distinguish between the motor element of a response and its releasing mechanism. When the motor responses in the experimental individuals raised in isolation are identical with those of normal controls, this does not necessarily mean that the releasing mechanism is the same. For instance, gulls and terns (like many other birds) feed their young; gulls by regurgitating and presenting small bits to their young, terns by presenting a freshly captured prey in the tip of the bill. These responses are innate. But their IRMs are changed by experience. During the first few days the parent birds are willing to feed any young of their own species, provided they are of the same age as their own young, but after several more days they have learned to know their young individually and respond to them alone; strangers are driven away (Tinbergen, 1936*b*; Watson, 1908; Watson and Lashley, 1915).

Whereas it is easy to see whether an individual raised in isolation shows the same motor responses as normal controls, it requires experimental study, and hence much time, to see whether the IRMs are the same in both. Thus it is only natural that we should know many cases of innate motor responses but only few of innate releasing mechanisms. Our knowledge of innate behaviour has been greatly widened by Heinroth, who has raised practically all the European bird species in isolation and studied their behaviour (see, especially, Heinroth and Heinroth, 1928). Lorenz has carried his work on and has extended it greatly (see Lorenz, 1931, 1935, 1941).

In older text-books another method for recognizing innate behaviour is often mentioned. If, it is said, the behaviour in all members of a species is alike, one can be pretty sure that it is innate. This, however, is a mistake. In numerous cases the external conditions under which the young grow up are in many respects exactly the same for all. In some species of songbirds, for instance, the nightingale, the song is individually learnt. As the young learn it from individuals of the same species, all of which have about the same song, every juvenile male individual learns the song of the species. Not only in cases like this, when the motor response is changed, but also in cases of change of an IRM by conditioning, the learning conditions may be the same for all individuals. As was stated first by Noble and Curtis (1939) and was corroborated by Baerends and Baerends (1950), some cichlid fish learn to confine their parental activities to young of their own species during the first time they breed. If a young pair are given eggs of another species in exchange for their own first brood, they will accept them and raise the young

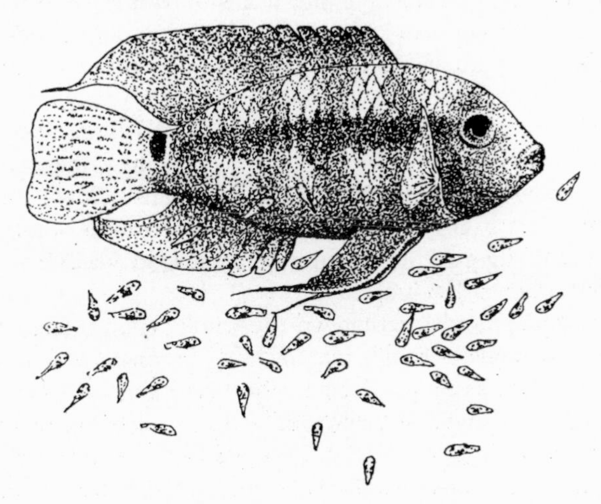

FIG. 50. Cichlid fish with young.

(Fig. 50). From then on they will never again raise young of their own species; they kill their own young as soon as they hatch. Normal experienced pairs will accept eggs of other species but they kill the young. This shows that, under normal conditions, these fish get conditioned to young of their own species when they breed for the first time. Here again, the fact that under natural conditions all individuals behave alike does not justify us in concluding that the behaviour is innate.

On the other hand, if it is observed that a certain response is not present in the young animal, this does not mean that it is acquired during individual life. First, a reaction may be innate and yet not appear before the animal is adult. The most extreme instances of such a state are the reproductive behaviour patterns. But, second, the gradual appearance of an activity during a slow, long period of development does not necessarily point to learning. The gradual improvement in the flying movements of birds, for instance, is only in part due to the acquisition of skill by practising. For the greater part it is the expression of a growth process, as will be discussed in Chapter VI.

In spite of all these pitfalls it is quite often possible to infer that an activity is innate without doing special experiments. As regards the change of IRMs by conditioning, there are a great number of species where the possibility does not present itself during normal life. For instance, how does a young cuckoo recognize and select a mate of its own kind? It has never seen a cuckoo before. And how could a male stickleback be conditioned to select only pregnant females of its own kind? From the moment of hatching it has only associated with young of its own age and with its father, and after it became independent it has only seen individuals in neutral condition, either males or non-pregnant females.

As regards the motor responses, there are even more species where learning by imitation is out of the question because they never get a chance to watch the performance of the response. Most insects, for instance, never have any contact with their parents' generation. Thus the complicated digging movements of a sphegid wasp can never be learned from other individuals.

Thus it is understandable that, on the whole, enough knowledge has been gathered to have a rough idea of what is innate and what is not. Reconsidering the examples given in the preceding paragraphs, it can be taken as certain that the behaviour of these sorts is innate: the reaction of birds to birds of prey, of the male three-spined stickleback to the female and vice versa, of the male robin to other robins, of *Eumenis* males to females, of insects to flowers, of a gull to eggs, &c.

What is 'a Reaction'?

So far we have concentrated our attention on the external causes releasing behaviour; we have called every movement released by an external stimulus 'a reaction' or 'a response'. It would be well to consider another element of a response, viz. the motor element, before continuing our study of the releasing agents.

In many cases the result of a stimulus is a very simple motor response. When we twitch the toe of a frog it simply withdraws the foot. When we touch the antenna of a locust it turns the antenna away. But when a dog walks through a herring gull colony in June, the gulls will utter their alarm call and the half-grown young react to this call by running to their shelters—each chick having a special hiding-place which it has learnt to use—and crouching. This is a more complex reaction, though still a relatively simple one. When 'unemployed' honey bees, waiting in the hive for a messenger, are at last activated by one performing the 'honey dance' (Fig. 51), the stimulus delivered by the dancer bee stimulates them to leave the hive. They fly in a definite direction over a definite distance (both communicated to them by the dancer) and begin to search for flowers, selecting only those that emanate the scent carried by the messenger. They suck honey, and after having made a 'locality study', they fly home. In this latter case the stimulus given by the messenger releases a complicated behaviour pattern (von Frisch, 1923, 1946).

These few examples may suffice to show that the concept 'response' or 'reaction' covers a wide variety of motor responses of very different degrees of complexity. Although it is quite justifiable to treat them as units as long as one is only concerned with the external releasing factors,

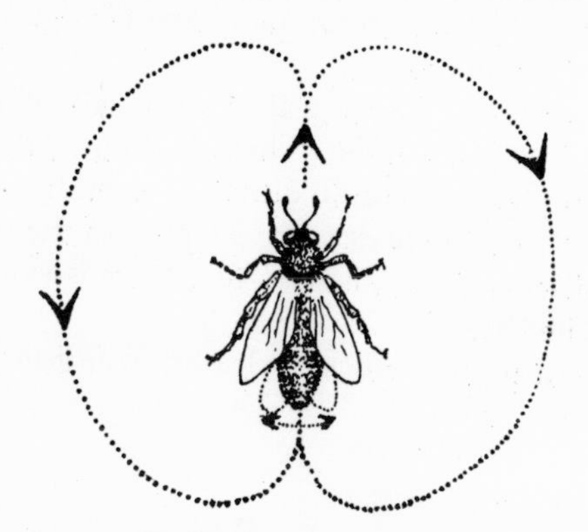

FIG. 51. The honey bee's dance. After von Frisch, 1946.

we should constantly bear in mind the fact that each of these 'units' is a more or less complex system, awaiting analysis. This analysis has already made some progress; it will be dealt with in Chapters IV and V.

Social Releasers

As we have already seen, many innate responses are dependent on stimuli given by other individuals of the same species. Because the study of these responses is of great importance for animal sociology, several have been investigated. The results confirm in a striking way the conclusion we have already drawn; many of these responses are dependent on the reaction of an innate releasing mechanism to a limited set of sign stimuli. Several instances mentioned in the paragraphs on sign stimuli and innate releasing mechanisms were concerned with such social responses.

The fact that these *social* responses provide the most striking examples of innate releasing mechanisms is not accidental. As we shall see later, the social relationships of many animals are based upon the functioning of structural or behavioural elements releasing specific responses in fellow members of the same species. These releasing features, whether movements or structures, sounds or scents have been singled out by Lorenz (1935), who called them *Auslöser* (releasers). This name has caused a good deal of confusion, because it seems to imply more than it does. A releaser in Lorenz's sense is not, in general, that part of an object the animal reacts to, but those features of *a fellow member of the same species* an animal reacts to. This limitation is absolutely essential, as we shall see later. I shall, therefore, translate *Auslöser* by 'social releaser' (Tinbergen, 1948*a*). Social releasers, therefore, are properties—either such of shape and or colour, or special movements, or sounds, or scents, &c.—serving to elicit a response in another individual, usually a fellow member of the same species.

Now the striking thing about social releasers is that they correspond exactly to the IRM they act upon. They send out little more than just the simple sign stimuli which are required to stimulate the corresponding IRM. It is as if social releasers are adapted to the properties of the IRM. As will be discussed in Chapter VII, there is evidence indicating that this is really the case.

The sociological aspect of social releasers will also be discussed in Chapter VII.

III

THE INTERNAL FACTORS RESPONSIBLE FOR THE 'SPONTANEITY' OF BEHAVIOUR

IN every study of the releasing value of sensory stimuli one is faced by the phenomenon of a varying threshold. The very same stimulus that releases a maximal reaction at one time may have no effect at all or may elicit a weak response at another time. This variation of threshold could be due to either (1) a variation of the intensity of another external stimulus not controlled in the experiment, or (2) a variation of the intensity of internal factors, or (3) both. In this chapter we shall consider the internal factors. The effect of these internal factors determines the 'motivation' of an animal, the activation of its instincts.

The methods of collecting facts bearing on this problem are of different kinds. First there are indirect methods. These are of three types: (*a*) changes of intensity or frequency of a response are observed under constant conditions; (*b*) the minimum intensity of the stimulus necessary to release a response is determined at different times while the conditions are kept constant in every other possible respect; (*c*) the minimum intensity of a stimulus required to inhibit a reaction is measured and its variations in the course of time are observed (obstruction method). The work done in these fields is rather fragmentary; nevertheless the results are of considerable interest.

Secondly there is more direct evidence. This has been obtained by studying the effects of experimentally controlled changes within the animal.

While the indirect evidence has been collected by students of behaviour, the more direct method was used by neurophysiologists and endocrinologists. The contact between these two types of investigators has not been what it should be; as a consequence too few attempts have been made to arrive at a coherent picture, although several tentative steps have been taken.

INDIRECT EVIDENCE

Variations of Intensity of Frequency of the Reaction under Constant Conditions

This phenomenon has been observed by many workers. However, very few careful and systematic studies have been made. Whitman (1919) summarized his extensive observations on the frequency of reproductive activities of pigeons in the course of the season in a 'frequency graph' (Fig. 52); the observed fluctuations were due partly to internal and partly to external factors.

Laven (1940) gives similar but more detailed graphs of the reproductive activities of the ringed plover (Fig. 53); here again, although no systematic analysis is presented, the fluctuations are doubtlessly due at least in part to internal conditions.

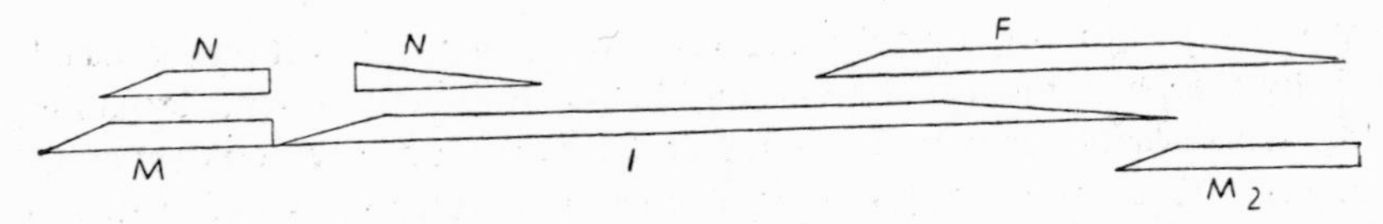

FIG. 52. Frequency graph of mating activities in pigeons. M, mating period; N, nesting activities; I, incubation of eggs and young; F, feeding of young; M_2, mating period of second cycle. After Whitman, 1919.

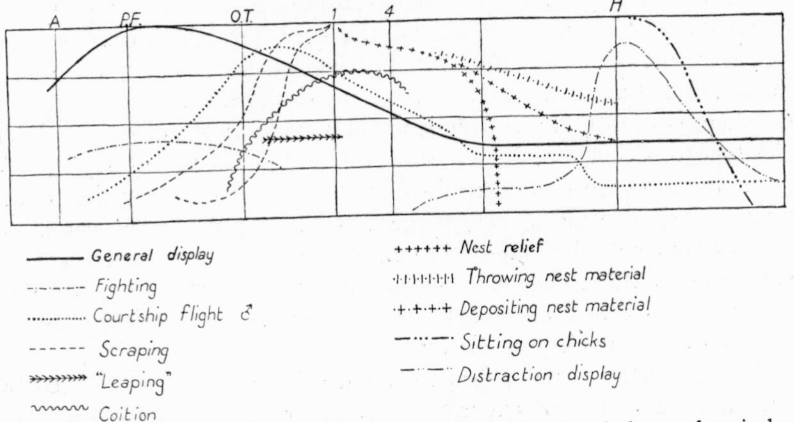

FIG. 53. Frequency graph of reproductive activities in the ringed plover. A, arrival on breeding grounds; $P.F.$, pair formation; $O.T.$, occupation of territory; 1, first egg laid; 4, fourth egg laid; H, young have hatched. After Laven, 1940.

Howard (1929) calls attention to the general phenomenon of the 'waxing and waning' of instinctive drives, without making clear to what extent waxing and waning are controlled by internal factors, although he is undoubtedly right in claiming that internal changes do play a part.

The migration drive in birds is another reaction showing wide fluctuations. The most extensive observations are those of Palmgren (1943); his records strongly suggest the influence of internal rhythms.

A more systematic approach has been made in a study of the three-spined stickleback. After a male has fertilized two or three clutches of eggs, its sex drive wanes and regular ventilation of the eggs begins. The time spent in ventilating ('fanning') increases daily until the eggs hatch, then fanning stops rather abruptly. The increase is partly due to an external stimulus: the increasing oxygen consumption of the eggs. Artificial lowering of the oxygen content of the water induces increased fanning. When half-developed eggs are replaced by fresh eggs, the fanning does not decrease but continues to rise until the day on which the original clutch of eggs preserved in another tank, hatches. The drop that follows then is not complete; the stimuli from the new clutch induce a secondary rise, culminating in a peak on the day on which the second clutch hatches. This has been repeated several times; each peak is lower than the preceding one, and as the external conditions are kept as constant as possible, and the stimuli from the eggs are of the same intensity in every cycle, the gradual decline must be due to an internal factor (van Iersel, personal communication) (Fig. 54).

Internal changes of another type have been observed in numerous other cases, of which the following is one. After a male stickleback has completed its nest, its nest-building drive remains active; the male continues to build. However, even under constant conditions, the building activities show marked fluctuations with periods of about 30–60 minutes (Fig. 55).

Repeated administration of the optimal stimulus required to release an instinctive response often results in a decreasing tendency to respond; the threshold becomes higher. This is especially obvious in those reactions that are but rarely released under natural conditions; the relative intensities of the 'impulse flows' from within going to the different motor mechanisms are nicely adapted to the biological needs.

For instance, 'injury-feigning' in a whitethroat, a reaction shown in response to a predator near the nest, can be released a few times in succession; the same stimulus that gave a strong reaction the first time fails to release it at the third or fourth repetition (Lorenz, 1937*b*).

As reported above, the sexual flight of the male grayling butterfly has been tested with dummies. When the same dummy was presented three times in short succession its releasing effect decreased; Fig. 56 gives the result of a great number of such series of three tests. The interval between two successive series was always more than 20 seconds. On the average, the releasing effect on the second test of each triplet was less than that on the first test, and that on the third test was still less. This is not the result of muscular fatigue; the reactions involved demand very little muscle activity in comparison with other acts of behaviour. It rather gives the impression of being due to exhaustion of the central nervous impulse supply, to central nervous fatigue.

Variation of Liminal Releasing Stimulus

Numerous observations, especially of the food drive and the sex drive, show that there is an enormous variation of the minimum stimulus required for the release of a reaction. Within certain limits, there is a correlation between the intensity of the liminal stimulus and the length of time that has elapsed after the last feeding or the last coition, as the

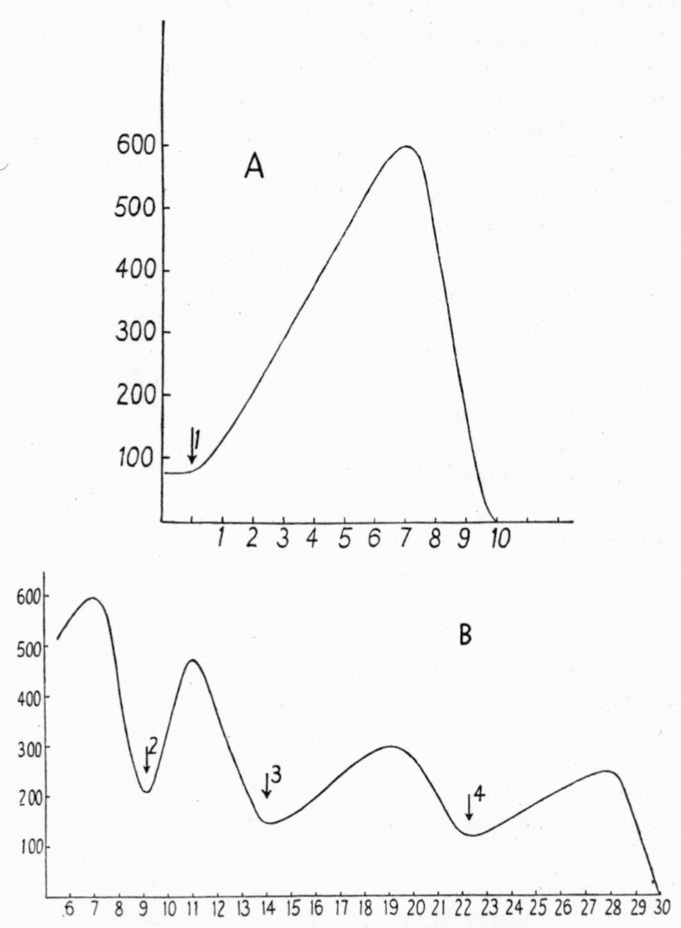

FIG. 54. Frequency of 'fanning' in male three-spined stickleback in course of time. Ordinate: seconds of fanning in sample periods of 30 minutes. Absciss: time in days. A: normal picture. Arrow 1 indicates moment of spawning. B: repeated peaks of decreasing height as a result of repeated replacement of eggs by younger eggs. Arrows 2, 3, and 4 indicate the start of new peaks each induced by fresh eggs. Courtesy of J. J. A. van Iersel.

case may be. It is not necessary to give many facts. Beach, in a review of this phenomenon in the sex drive in male mammals, concludes that 'The specificity of the stimulus adequate to elicit mating responses varies inversely with the sexual excitability of the individual' (1942, p. 174).

It is necessary, however, to point out that this conclusion is too often taken for granted without realizing the relatively poor evidence upon which it is based.

We know that, for example, long deprivation of food lowers the threshold of responses to food considerably, and that males in sexual condition become more and more ready to mate with suboptimal substitutes of females when they are not allowed to mate with an optimal female of their own species.

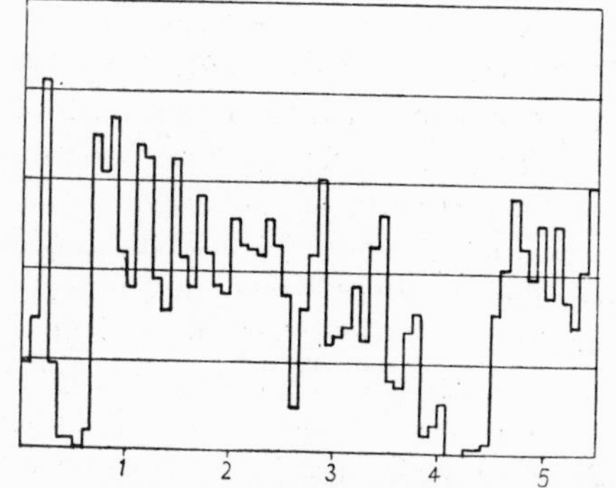

Fig. 55. Frequency graph of building in male three-spined stickleback under constant external conditions. After Tinbergen, 1948.

In many cases, however, changes in readiness to respond are taken without further proof as indications of changing motivation, and the relations between them are presupposed rather than inferred.

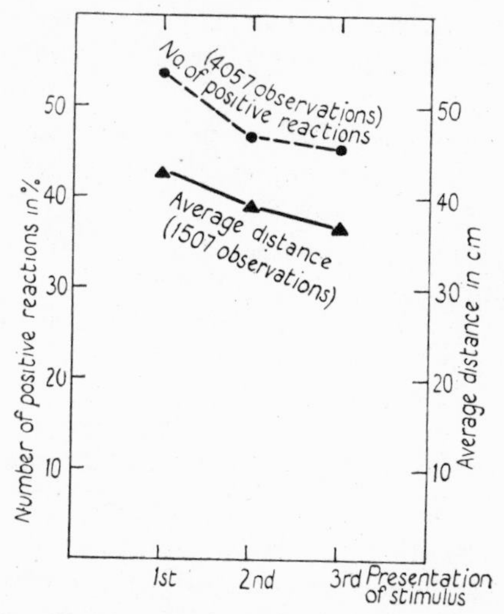

Fig. 56. 'Central nervous fatigue' of sexual pursuit in male grayling as shown by decrease in responsiveness to repeated presentation of model. After Tinbergen, Meeuse, Boerema, and Varossieau, 1942.

Although there is no reason to doubt the general validity of this assumption, yet careful studies in which both the liminal external stimulation and the internal motivation are actually controlled and measured are greatly needed. The most careful study of this kind is perhaps Hemmingsen's work (1933) on the sexual behaviour of the female rat. In the main, it confirms our conclusion. Similarly, the experiments by Wiesner and Sheard (1933) on the maternal behaviour pattern of the female rat confirm that there is a correlation between internal factors (hormone level) and responsiveness to external stimuli.

A drive may even become so strong that its motor responses break through in the absence of a releasing stimulus. Lorenz (1937*b*) observed these 'explosions' or 'vacuum activities' in a captive starling. It performed repeatedly the whole behaviour pattern of insect hunting, from watching the prey through catching, killing, and swallowing, without any discernible external stimulus; Lorenz made sure that no insects, however small, were releasing this complex pattern. Critics (e.g. Bierens de Haan, 1937) have pointed out that one can never be sure that there has not been a slight stimulus such as a floating dust particle. This may be correct, but even so, this observation is proof of an extreme lowering of the threshold of the response. Recently more instances of vacuum activities have been published. Kluyver (1942) observed numerous cases of vacuum activities of about the same type of insect-hunting behaviour in European waxwings during cold frosty weather when he could not see any prey. Young honey buzzards perform the typical movements of digging out a wasps' nest while still in the nest (Gentz, 1935); many birds sit in the nest before they have laid eggs (Howard, 1929; Verwey, 1930). It is a remarkable fact that even the escape reactions of some birds have been observed to go off *in vacuo*. Lebret (personal communication) has watched several species of ducks going through the specialized movements of escaping a charging duck hawk (diving from considerable height, swimming under water with the use of the wings, following a zigzag course) in the absence of any predator. Krätzig (1940) raised two broods of ptarmigan, one without any adult birds, the other with a foster mother. In the isolated birds escape was never released in the natural way (i.e., by the alarm call of the mother) and they developed an increasing tendency to show frantic flight responses at the least disturbance (e.g., a wasp) or even without any discernible external stimulus.

I have myself observed several instances of vacuum courtship and other reproductive activities in sticklebacks; for instance, a complete 'zigzag dance' in an empty tank.

Change of Liminal Inhibiting Stimulus

Instead of studying the strength of a stimulus necessary to release a reaction it is also possible to study the inhibitory influence of obstacles blocking the reaction. By confronting an animal that is striving towards the accomplishment of an instinctive act (feeding, mating, nursing young, &c.) with a measurable obstruction of varying strength, it is possible to determine, and to express in a quantitative way, the degree of obstruction required for the inhibition of the act. The greater the obstruction tolerated, the stronger the motivation must be. This method has been used to measure the strength of a drive at different times, to compare the strength of a drive in different individuals, and also to compare the strength of different drives in one and the same individual. The value of the latter procedure is highly problematical, for who could judge whether the incentives offered for the release of different drives, e.g. food or a receptive female, are of equal strength? However, as a means of measuring the strength of one drive at different times or in different individuals, the method is useful, though perhaps unduly elaborate.

Nice (1937) reports a 'natural experiment' of this type. The singing of male songbirds is a reproductive activity. In many birds it can be inhibited by low temperatures. Nice measured the temperatures that inhibited song in male song sparrows at various days throughout early

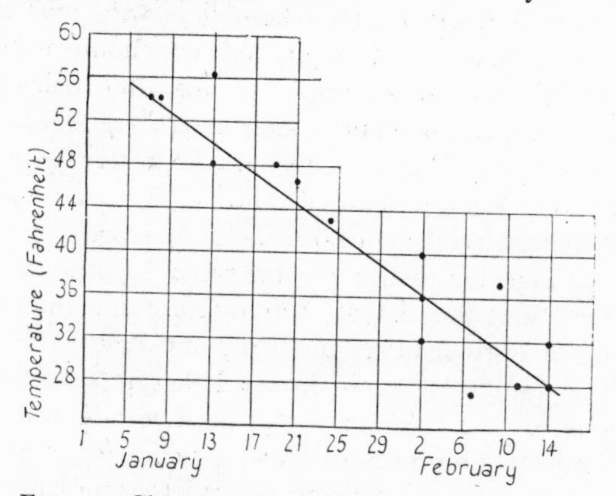

FIG. 57. Changing threshold for temperature as a stimulus suppressing song in the song sparrow (*Melospiza melodia*). After Nice, 1937.

spring. It was found that the liminal temperature became lower as the season advanced (Fig. 57), which can only be due to a gradual increase of the motivational factors.

Although all this evidence is both fragmentary and indirect, in its diversity it suggests the conclusion that many reactions are not pure responses to external stimuli but are dependent on internal factors as well. We will now examine the nature of these internal factors more closely.

DIRECT EVIDENCE

Hormones

Not long ago endocrinologists rarely studied the influence of hormones on behaviour; their interest was focused primarily on problems of growth. Only occasionally did they mention behaviour elements, and if so, they used them as more or less accidental indications supporting their primary evidence. In recent times, however, behaviour has been included in several studies and it is certain that some of the internal factors responsible for the fluctuations in responsiveness and for vacuum activities are hormones. This is especially obvious in reproductive activities. In others, like reactions to food or escape reactions, it is less probable that hormones are the internal factors responsible.

The evidence concerning the influence of sex hormones on behaviour, though sufficient to justify the conclusion that such influence is far-reaching, is still rather fragmentary. Of course, many indications are found in the fact that castrated animals do not show complete reproductive behaviour, but further analysis is only just beginning. An exhaustive review has been given by Beach (1948).

In the American chameleon, *Anolis carolinensis*, the male sex hormone testosterone propionate, when implanted subcutaneously as pellets into gonadectomized and intact immature and adult males and females, induces several elements of male sexual behaviour, such as territorial fighting and copulation. However, in both sexes female behaviour was induced in the very same individuals, although it was very incomplete in the treated males (Noble and Greenberg, 1940).

Evans (1935) injected males of the same species in winter with pituitary extracts and found that their fighting and mating behaviour resembled that of normal males in spring and early summer, whereas the controls did not change their normal lethargic winter behaviour.

In birds most of the work has been done with the domestic fowl. Male chicks injected with testosterone propionate exhibited all the sexual behaviour patterns (crowing, complete copulatory behaviour) of the adult cock (Noble and Zitrin, 1942). A female chick, after receiving a pellet of testosterone propionate when 5 months old, copulated in the male fashion after $4\frac{1}{2}$ months. On one occasion it crowed three times (Zitrin, 1942).

Male valley quails entered the sexual cycle much earlier than normally after implantation of testosterone propionate pellets (Emlen and Lorenz, 1942). Boss (1943) induced premature sexual behaviour in herring gulls by injections of testosterone propionate. Davis and Domm (1943) observed crowing, 'waltzing', 'tidbitting', and copulation in two of three capons after injection with testosterone propionate. Three bilaterally ovariectomized poulards reacted by crowing, waltzing, and fighting but did not copulate.

Rowan (1932) subjected crows to artificially lengthened days in winter. The result was growth of the testes and at least in some cases northward migration, which suggests that the spring migration, the first element of reproductive behaviour, is activated by either pituitary or gonadal hormones. For a full discussion of the problems awaiting solution here see Wolfson (1941, 1942, 1945).

In female canaries song and some elements of male courtship behaviour have been induced by administration of testosterone propionate (Frederiks, 1941; Shoemaker, 1939).

In the night heron both sexes, when treated as immature birds with testosterone propionate, develop the complete masculine courtship and mating pattern (Noble and Wurm, 1940).

In male pigeons castration usually abolishes sexual activity. In some of the birds, however, complete coition could still be released by very active females (Carpenter, 1933). This shows once more the relation between internal and external factors: even in the absence of a very important internal factor, administration of optimal external stimuli may release the reaction.

The influence of œstrogens on the female behaviour patterns has likewise been studied, but the results are more fragmentary. Poulards receiving oestrogen squatted for the rooster. In capons œstrogen induced male copulatory behaviour but no crowing or 'tidbitting' (Davis and Domm, 1943). Female chicks, receiving daily injections of œstrogen starting on the 15th day of age, squatted for treading males after about three weeks (Noble and Zitrin, 1942).

Work on mammals is mostly confined to the rat, but some work has also been done with other species, e.g. with rabbits, with dogs, and with man.

In mammals prepuberal castration prevents males from developing the reproductive pattern, post-puberal castration effects a gradual loss of sexual activity. Administration of androgen in all such cases restores sexual behaviour. In mammals œstrogen also activates the male's sexual behaviour.

Sexual behaviour of female mammals is also dependent on hormones. Spayed female rats do not exhibit receptive behaviour to males. After treatment with œstrogen and progesteron combined, sexual activity becomes normal again (Beach, 1943, 1948). Essentially the same has been found in *Cavia* (Dempsey, 1939).

The evidence thus far obtained shows that gonadal hormones are an important internal factor of reproductive behaviour. However, the facts that (1) capons and poulards still behave differently after the same hormone treatment, and (2) complete castration of immature male pigeons may not entirely prevent the development of complete copulatory behaviour, suggest that gonadal hormones are not the only internal factors. Either other hormones or intrinsic central nervous factors must play a part too. The curious fact reported by Beach (1948) that female rats possess the mechanism of the complete masculine mating pattern and that this behaviour is not influenced by ovariectomy points to the importance of intrinsic central nervous factors. In this connexion it is also of considerable interest that œstrogen may activate male behaviour in males and female behaviour in females.

A hormone influencing the maternal phase of the reproductive cycle is prolactin. In pigeons prolactin activates the parental behaviour pattern as well as crop secretion. In fowls treatment with prolactin induces broodiness in laying hens, the intensity of the reaction depending on innate capacities: broody races respond better than bad brooders like the white leghorn (Riddle, 1941). Domestic cocks injected with prolactin react to chicks by brooding them; they lead them to the food, utter alarm calls as a reaction to a predator, and defend the young. In other words, they show the complete maternal behaviour pattern typical of the broody hen. However, even strong doses of prolactin did not bring them to sit on eggs (Nalbandov and Card, 1945).

Thus it is obvious that in reproductive behaviour at least hormones play an important part in determining the motivation. The problem of how they influence behaviour will be discussed below. It may be stated in advance that all the evidence points to the conclusion that the hormones act directly upon the central nervous mechanisms.

In this connexion, one fact may be discussed here. Many observations show that a low hormone level does not have the same effect on all the elements of the reproductive pattern. Some elements, like crowing and waltzing in the domestic cock, require a less intense hormone stimulation than others, such as actual copulation. The obvious conclusion would seem to be that the former elements have a lower threshold for the hormone. However, as we must assume that the hormone does not act directly on each of these elements of the pattern, but through central nervous mechanisms superordinate to the separate elements, the loss of integration caused by a lowering of the hormone level is probably a central nervous affair depending on the unequal distribution of impulses flowing from the central mechanisms to its subordinated elements. It is not the latter's reactions to the hormone that have a different threshold, but their reaction to the impulse from the superordinate centre.

Internal Sensory Stimuli

The best known example of internal sensory stimuli playing a part in the regulation of behaviour is that of the 'hunger drive'. It is generally known that the responsiveness to food increases and that spontaneous searching for food is stimulated by the internal condition perceived, in man, as 'hunger'. Hunger is a complicated phenomenon, but part of it at least is dependent on stimulation by the contractions of the muscles of the stomach wall. These contractions occur in rhythmic sequence and give rise to the 'hunger pangs'. This is not the only internal source of the hunger drive, for removal of the contractile part of the stomach in rats had only a very slight effect on hunger motivation (Tsang, 1938).

Intrinsic Central Nervous Factors

Apart from hormones and internal sensory stimuli there may be a third type of internal factor. Indirect evidence for its existence is of several different types. First, there are observations of spontaneous behaviour that cannot be explained as effects of either hormones or internal stimuli. Domestic dogs, for instance, start on hunting excursions under conditions indicating that stomach contractions are playing no part; the hunting behaviour cannot always be suppressed by abundant feeding. Second, the observations on specific fatigue or exhaustion of responsiveness, described above, suggest the activity of a slowly accumulating internal factor that needs time for restoration after each response.

Third, many vacuum activities, such as the escape reactions observed by Krätzig (1940) and by Lebret, suggest that hormones and internal stimuli are not the only internal factors.

Lorenz (1937*b*), who was the first to point out the existence of vacuum activities, supposed that the central nervous system itself produced impulses that were acting as specific causes of instinctive patterns.

Several neurophysiologists have recently attempted to study the problem of spontaneity in a more direct way. Of especial importance for ethology are the studies concerning rather complicated co-ordinations. The highest level of integration that has been studied so far is that of locomotion. After a period in which the attention of neurophysiologists was focused on reflex phenomena, which led to a certain neglect of other aspects of nerve activity, Adrian and Buytendijk's discovery (1931) of intrinsic activity of the respiratory centre of the goldfish was the beginning of a period of increased interest in possible intrinsic or 'automatic' functions of the central nervous system. Since then a number of studies, of invertebrates as well as vertebrates, have shown that the classic conception of locomotion as reflex chains and patterns of essentially reflex character was one-sided.

Von Holst (1934) found that the locomotory movements of the two pectoral fins and the caudal fins of the goldfish under certain conditions of anaesthesia become exactly synchronous with the respiratory movements. This he considered as an argument in favour of the automatic nature of the nervous mechanisms underlying the locomotory rhythm. Further studies showed that we must distinguish between (1) the initiation of the movements, (2) the rhythmical contractions of the muscle group of each somite, and (3) the propagation of the waves of contraction along the trunk, or along the rays of separate fins.

Initiation. Opinions differ about the initiation of the movements. One way to settle the question is to see whether complete deafferentation prevents locomotory movements. The results obtained so far (von Holst, 1935, in the teleost *Tinca vulgaris*; Lissmann, 1946, in the elasmobranchs *Scyllium canicula* and *Acanthias vulgaris*; Gray and Lissmann, 1940, in the toad *Bufo bufo*) show that complete deafferentation abolishes the locomotory movements.

Weiss, however, reports (1941*c*) that complete deafferentation in tadpoles did not impair swimming. Further, Ten Cate (1939), von Holst (1935), and Lissmann (1946) report that the complete locomotory pattern appears as a consequence of a minimum of exteroceptive inflow. Spinal preparations of the tench (*Tinca*) in which all dorsal roots except the two innervating the pectoral fin were severed, spinal sharks (*Scyllium*) in which the dorsal roots except the three innervating the pectorals were destroyed, and spinal sharks (*Scyllium* and *Acanthias*) in which either the dorsal roots of the anterior half or those of the posterior half, or those of either left or right side, were severed showed complete locomotory movements of slightly submaximal intensity (Fig. 58). Similarly, a tench in which all dorsal roots except the two innervating the pectorals were severed while the connexion with the brain was left intact could be stimulated to swim by a wide variety of stimuli of tactile, static, visual, or nociceptive nature (von Holst, 1935).

The initiation of these movements, therefore, would not seem to be of an absolutely automatic, purely intrinsic nature, but the stimulus required is (1) very slight and (2) highly unspecific.

At a first glance, this seems more or less contradictory to the following facts. In the isolated ventral cord of the beetle *Dytiscus* (Adrian, 1931) and of the annelid worm *Lumbricus* (von Holst, 1937) rhythmic activity could be traced by the action current method, just as in the respiratory centre of the goldfish. Von Holst concludes that these rhythmic waves of *Lumbricus* corresponded to the locomotory movements.

An experiment of quite another type, carried out by Weiss (1941*b*), also points to independent activity of the spinal cord. In experiments primarily designed for the study of the growth of nerve-fibres, Weiss implanted in an intact axolotl a piece of embryonic spinal cord in the connective tissue of the dorsal trunk musculature. At the same time, a fore limb *anlage* was implanted (Fig. 59). The implanted nerve tissue did not grow a connexion with the central nervous system of the host but grew motor and sensory fibres towards the grafted limb. The motor neurones were connected with the limb muscle several weeks before the sensory fibres had come into contact with the receptors in the graft. Yet, as soon as the motor neurones had established contact, the limb began to carry out rhythmic movements. This means that the implanted spinal cord was responsible for the production of impulses independent of any external stimulation.

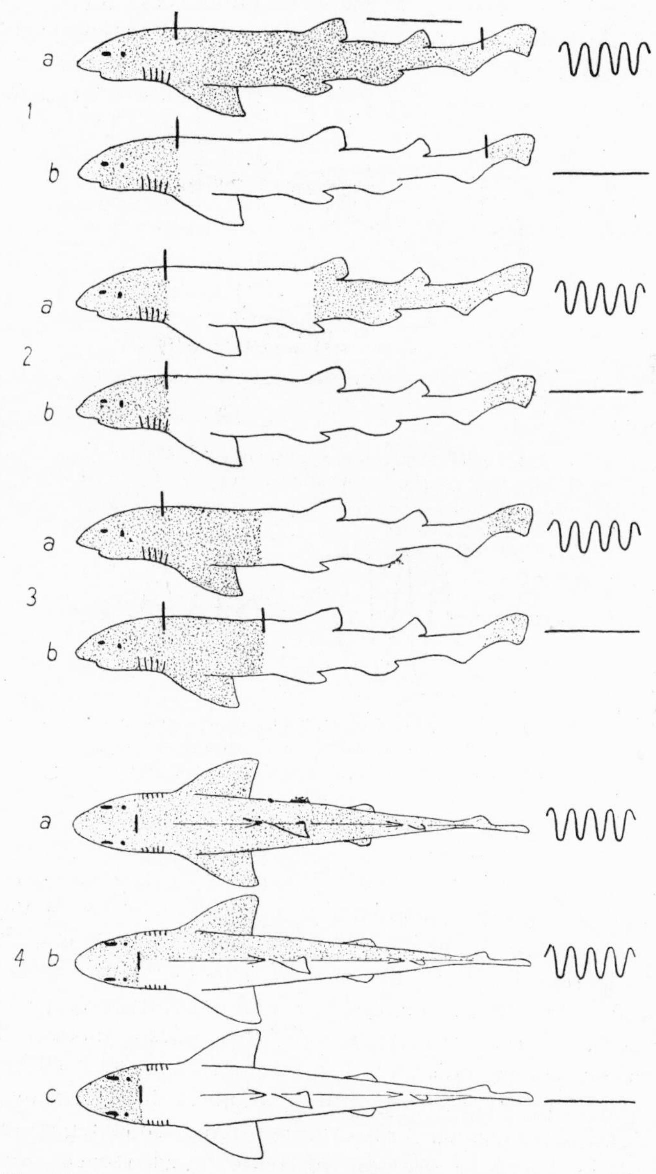

FIG. 58. Locomotory movements (as registered on kymograph, right) of sharks after local deafferentation. Shaded: dorsal roots intact. White: dorsal roots severed. Crossbars indicate severance of spinal cord. After Lissmann, 1946.

The facts at our disposal are, therefore, conflicting. It seems that they could be reconciled in either one of two ways. Either (1) the unspecific inflow raises in a direct way the general excitatory state, or (2) the inflow removes a block that prevented the automatic impulses from expressing themselves in motion. While some of the available evidence seems to support the first hypothesis, the following facts point to the second. In

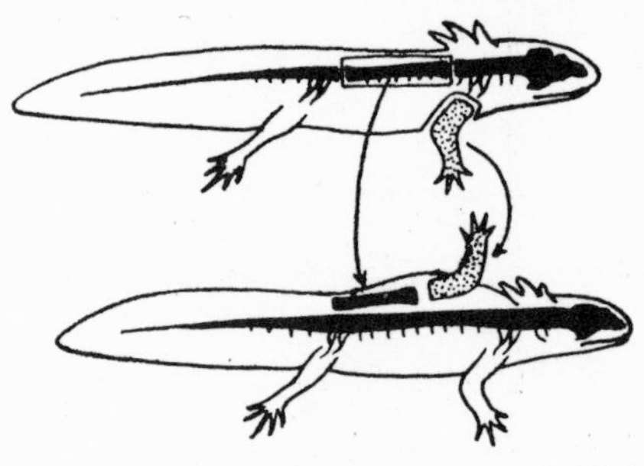

FIG. 59. Deplantation experiment by P. Weiss (1941). Explanation in text.

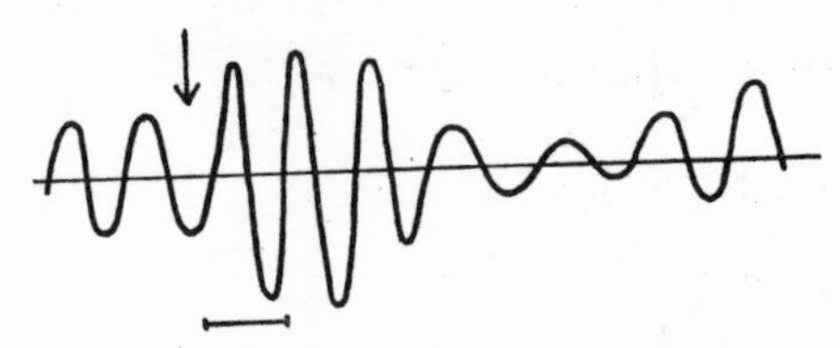

FIG. 60. 'Negative after-effect' of external stimulation on intensity of swimming movements in goldfish. Arrow indicates beginning of stimulation; duration of stimulus indicated by horizontal line. After von Holst, 1934.

the goldfish, rhythmic trunk movements carried out by a preparation in which the medulla is severed just in front of the vagus can be increased by directing a weak water current to the skin of the trunk. After this period of increased response the movement does not return to its former intensity but decreases below this level and only increases again after a certain time (von Holst, 1934) (Fig. 60). This indicates that the stimulus does not increase the excitation of the automatic centre itself but gives it the opportunity of discharging more impulses than normally.

A goldfish whose spinal cord is severed in the caudal region responds to stimulation by a local water current by one single movement towards the source of stimulation, which is the simplest element of the response pattern of the trunk. When the external stimulus is repeated a number of times in quick succession, the response decreases in intensity. This phenomenon of fatigue also points to exhaustion of the motor impulse flow.

The evidence on locomotory patterns thus far obtained therefore seems to point to intrinsic activity of the spinal cord and medulla. Some of the external stimuli seem to be necessary to raise this activity so as to bring it above threshold value; other stimuli obviously remove a block and thereby provide free passage to the motor impulses from the automatic centres.

Of course this does not mean that reflexes are not involved; however, it means that they have not the overriding importance at one time attributed to them.

These facts are of paramount importance for our understanding of innate behaviour. So far they point to a close parallel between the mechanism of the innate patterns as a whole and the locomotory patterns which are their constituent parts. It seems that these studies have opened the way for neurophysiological study of the 'spontaneity' of behaviour.

Rhythmic activity of a somite. The rhythm within each somite, causing alternating contractions of left and right longitudinal muscles, has long been considered a chain reflex. Yet there is proof that the rhythm is at least in part controlled by intrinsic activity of the spinal cord which, though normally reinforced by proprioceptive reflexes, could keep up the rhythm after severance of the afferent nerves. On this foundation, Graham Brown as early as 1912 had formulated a theory of the automatic nature of quadruped ambulation. The extensive investigations of von Holst on the locomotion of fishes (summary in von Holst, 1937) and also the work of Gray, Ten Cate, Lissmann, and others showed that both in elasmobranch and teleost fish and in amphibians the alternating rhythm of single somites or of pairs of fins or legs is not impaired by deafferentation.

For our present purpose it must be sufficient merely to stress this fact without further comment; the nature of the integrative principles involved will be discussed in Chapter V.

Propagation along the trunk.—In addition to rhythmical movements of each somite or pair of limbs, normal locomotion involves propagation of the contractions along the body axis. This co-operation between successive somites results in sinusoid undulatory movements of the trunk (eels and other fishes) and, on a lower level, in sinusoid undulation of separate fins or in alternating rhythms in fore and hind limbs. This co-ordination has been considered a reflex chain ever since Friedländer (1894) showed that in the earthworm the contractions of each somite were dependent on proprioceptive stimulation caused by the pull of its predecessor. In this or other ways, propagation of a contraction wave was judged to be dependent on proprioceptive stimuli originating either in stretched skin or in contracting muscles. These stimuli would activate, in reflex fashion, the contraction in the whole series of muscles involved.

If this were the only mechanism responsible, deafferentation should disturb the co-ordination. Experiments by von Holst (1935), Gray

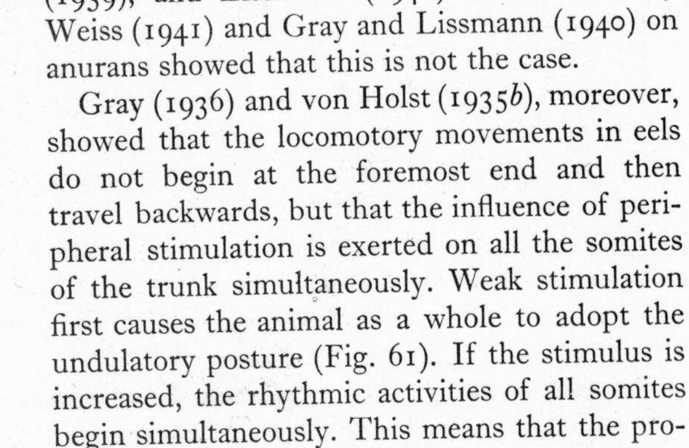

(1939), and Lissmann (1946) on fish and by Weiss (1941) and Gray and Lissmann (1940) on anurans showed that this is not the case.

Gray (1936) and von Holst (1935*b*), moreover, showed that the locomotory movements in eels do not begin at the foremost end and then travel backwards, but that the influence of peripheral stimulation is exerted on all the somites of the trunk simultaneously. Weak stimulation first causes the animal as a whole to adopt the undulatory posture (Fig. 61). If the stimulus is increased, the rhythmic activities of all somites begin simultaneously. This means that the process should not be considered as a propagation of impulses along the trunk, but of simultaneous rhythmic activity of the whole series of somites.

There is a very definite relation between the phases of these rhythms, resulting in the propagation of the trunk undulations in the caudal direction. The co-ordination of the phases is present right at the start.

A discussion of the mechanism of this co-ordinated functioning must here again be postponed to Chapter V.

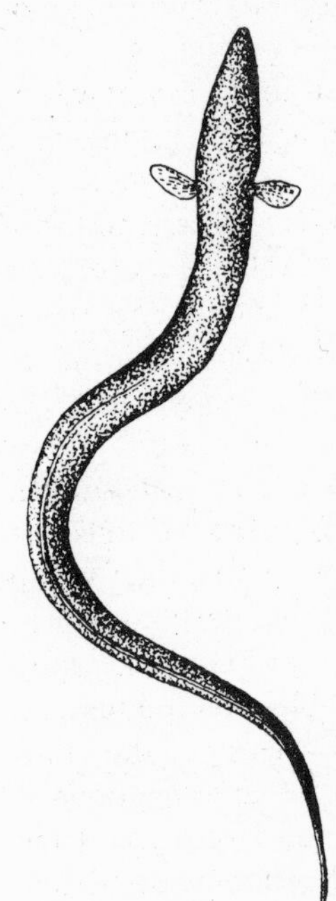

FIG. 61. Undulatory posture of eel (*Anguilla* L.)

In summary, the work on locomotion shows an interesting parallel to that on instinctive activity as a whole. It is probably not without significance that, although the work in the two fields was done independently, it led in both cases to a rejection of a pure reaction theory. In both fields a more balanced view is developing in which there is room for both spontaneity and reactivity. Moreover, in both fields the problem of spontaneity has now been made accessible to objective causal research, because it is now realized that internal factors are not necessarily of a purely subjective order. This view was the outcome of a confusion of 'spontaneity' with the essentially subjective concept of 'free will'. The cause for the neglect of internal factors is found in the greater technical difficulties encountered in attacking them.

A further step in the analysis was taken by von Holst (1935*b*). In discussing his crucial experiment we must consider in advance a phenomenon that will be discussed more fully below. Co-ordination between the rhythms of the pectoral fins and that of the dorsal fin is partly due to a process called 'superposition'. The pectoral rhythm, which is the so-called dominant rhythm, is superimposed upon the rhythm of the dorsal fin, which is therefore called the dependent rhythm (Fig. 62). As a result, the movements of the caudal fin appear as a complex rhythm. This process of superposition is independent of peripheral stimuli, for (1) in deafferented spinal fish superposition is unimpaired, and (2) pas-

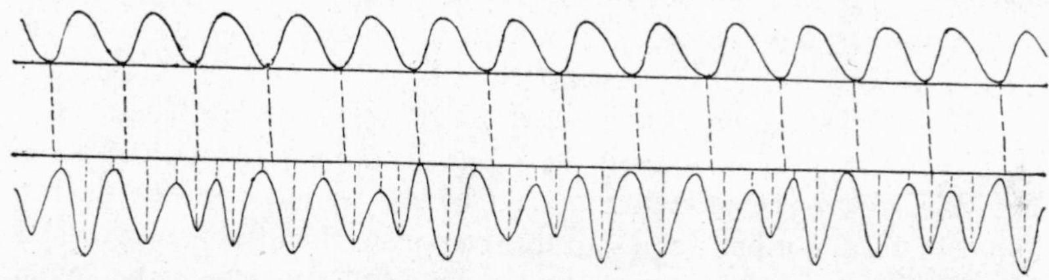

FIG. 62. The 'superposition effect'. Registration of the movements of the left pectoral fin (dominant, above) and of the dorsal fin (dependent, below) of *Labrus*. Amplitude of dependent fin increases when tops of the two are reciprocal; decreases when tops coincide. Slight irregularities in tempo of dependent rhythm are due to the magnet effect (see Fig. 95). After von Holst, 1935*c*.

sive movements of the pectorals have no influence on the dependent fin. This shows that the movements of the pectorals are controlled by motor centres which in their turn receive impulses from other centres.

THE CO-OPERATION OF THE CAUSAL FACTORS

After having thus reviewed the various causal factors, external and internal, that are responsible for an innate reaction, our next task is to consider the problem of their co-operation. Although a full discussion is not possible before we have studied other aspects of the causal organization of behaviour, some partial problems must be considered now.

One thing is already evident: there is a mutual relationship between internal and external factors in the sense of an additive influence on the motor response. A high intensity of one factor lowers the threshold for the other factors. A high hormone level increases the responsiveness to external stimulation; if the hormone level is low, very intensive external stimulation is required to bring the total of causal factors above threshold value. It is evident that, while the impulses derived from sensory nerves are qualitatively different from those coming from the other sources, their influence on the motor response is added quantitatively. As we shall see later (Chapter V), there is more evidence pointing in the same direction, viz. that each 'reaction' is controlled by its own special mechanism in the central nervous system.

Special consideration is due to the part played by hormones. Lashley (1938), in an important review of the analysis of instinctive behaviour, has discussed four possibilities: (1) Does the hormone stimulate the growth of new nervous connexions? (2) Does the hormone act merely by increasing the general excitability of the organism? (3) Does the hormone act by inducing specific changes in various organs, which in their turn initiate sensory impulses which facilitate the mechanisms of special reactions? (4) Does the hormone act upon the central nervous system, thereby increasing the excitability of the sensorimotor mechanism specifically involved in the instinctive activity?

The first and the third possibilities are ruled out by the occurrence of quick responses to changes in hormone level. For instance, Wiesner and Sheard (1933), studying the retrieving response of female rats, found that the normal waning of this response when the young reached weaning age could be revived under conditions in which the organs that could be supposed to act as intermediates were absent. Thus, Moore (Lashley, 1938) interchanged gonads between male and female rats and noticed retrieving behaviour in the feminized males.

The second possibility, that of general increase of excitability, does not cover the facts at all. As already emphasized in Chapter II, it is not specific muscles that are engaged in special instinctive responses, but rather specific patterns of muscular activity. Two different responses may use the same muscles, but their contractions are differently integrated. When, as is the observed fact, hormones selectively activate specific patterns, this cannot be explained by an hypothesis of a general, but only of a specific, rise in excitability, specific in the sense of applying to special motor centres. Moreover, we actually do know what happens when the general excitability is raised: general tetanus, resulting in chaos, is the result of administration of strychnine, which is such a general excitant.

It is obvious, therefore, that we must consider the fourth hypothesis as the most probable one.

CONCLUSION

The schematic review presented in this chapter shows two things clearly. First, we may now draw the conclusion that the causation of behaviour is immensely more complex than was assumed in the generalizations of the past. A number of internal and external factors act upon complex central nervous structures. Second, it will be obvious that the facts at our disposal are very fragmentary indeed. Some of our conclusions apply to instinctive patterns as a whole, others apply to the locomotory constituents only.

We should be able, however, to construct a tentative hypothesis about the system of initiating factors. This picture can only be provisional and we will have to extend it and make it concrete in the next chapter.

We have seen that there are close parallels between the mechanisms underlying locomotion and those underlying an instinctive act as a whole. Both are subject to internal factors and external stimuli. Both show the tendency to 'explode' in the absence of releasing external stimuli. These facts have led Lorenz to put forward the hypothesis that instinctive responses too are controlled by automatic centres that send out a continuous flow of impulses to central nervous motor mechanisms. Some kind of block prevents discharge into muscle action, which would lead to chaos. Discharge is brought about by adequate stimuli, namely, the combination of sign stimuli typical of each instinctive response. These sign stimuli act upon a reflex mechanism, the innate releasing mechanism, which alone is able to remove the block, thus allowing the accumulated impulses to discharge themselves in muscle actions constituting the motor response.

Von Holst has characterized this situation by stating that each automatism is enveloped in a 'coat of reflexes'.

> The nervous system, therefore, is not like a lazy ass that has to be beaten or rather has to pinch its own tail before being able to do one step, but rather like a temperamental horse that has to be kept in check by the reins and urged by the whip (von Holst, 1937).

IV

FURTHER CONSIDERATION OF THE EXTERNAL STIMULI

SIGN STIMULI

THE term 'stimulus' is, in a sense, misleading, in spite of its apparent clarity. It suggests that we have to do with a unit of energy which is open to quantitative study and can easily be measured. In reflexology, it is true, the stimulus applied is often simple: a measurable electric shock, or a certain amount or concentration of a chemical compound, &c. A close study of sign stimuli, that is of the complete stimuli to which an animal's instinctive movements respond (as opposed to the artificial and often partial stimuli applied by the laboratory worker), shows that they are, in reality, very complex and not readily measurable. Let us consider some instances.

Nestling passerine birds respond to the parents' return with food by gaping. This response, which is innate, has been studied in the thrush *Turdus merula*. The very young nestlings are still blind and their gaping response is released by tactile stimuli. At the age of about 8 days, when the eyes are open, visual stimuli release the response. At about the same time, the direction of the gaping response changes; whereas in the blind stage the neck was stretched vertically, it is now directed towards the parent's head.

Tests with models showed that one of the sign stimuli characterizing a head was size. Shape was of little importance: any external interruption of the body's outline directed the gaping towards it. When a model with two heads of different size was presented, the birds showed distinct preference for one of the two heads. Thus in the model represented in Fig. 63*b*, the larger of the two heads was reacted to. However, in the dummy of Fig. 63*a*, which has exactly the same heads but a much smaller body, the smaller head was preferred. The sign stimulus, therefore, is not a head of a certain absolute size, but one of a certain proportion to the size of the body. Thus the 'stimulus' is a clear example of a 'configurational stimulus' (Tinbergen and Kuenen, 1939).

Another instance was found in the food-begging response of the herring gull chick, which, as was shown on p. 30, is a reaction to the red patch of the parent bird's lower mandible. Two models were presented to chicks, in which the location of the red patch was varied. The relative frequency of pecking responses released by each of these models is represented in Fig. 64. The results lead to the conclusion that the releasing value of model *b* is only one-fourth of that of model *a*. Although in model *b* all the elements of the standard model are present, their spatial arrangement is different. Clearly, the sign stimulus is not simply 'something red', but 'a (preferably red) patch at the tip of the lower mandible'. The location of the red patch in relation to other parts of the head is important. We are again dealing with a configurational stimulus (Tinbergen, 1948*b*, 1949).

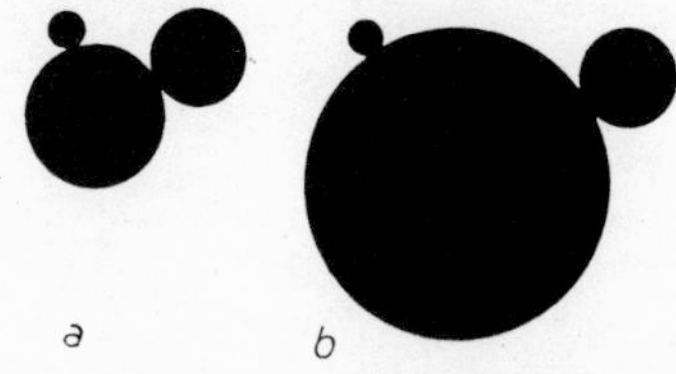

FIG. 63. Models with two heads used to release gaping responses in nestling thrushes. Explanation in text. After Tinbergen and Kuenen, 1939.

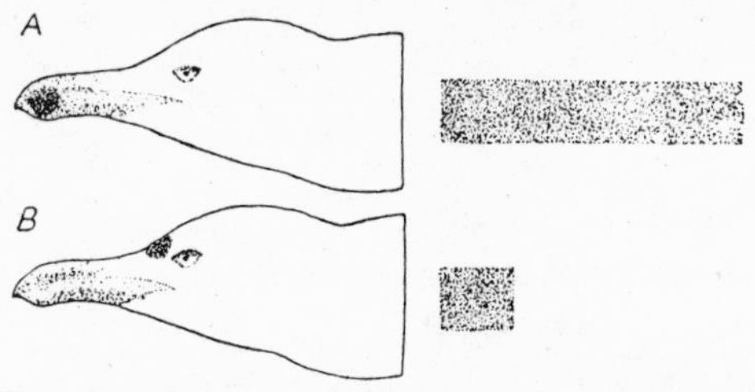

FIG. 64. Models of herring gulls with different location of red patch. Columns indicate their value in releasing the chicks' begging response. After Tinbergen, 1948.

The reactions of young gallinaceous birds, ducks, and geese to a flying bird of prey are released by the sign-stimulus 'short neck' amongst others (see p. 31). The amazing complexity of this stimulus is shown by the following facts. A model was made with symmetrical anterior and posterior wing edges (Fig. 65). At one end of the body axis there was a short protuberance, at the other end a long one. When this model was sailed to the right it had a short neck and a long tail. When flown in the other direction the neck was long and the tail short. In the first case it elicited escape responses; in the latter it did not. As this difference cannot be based upon wing shape, it must be dependent on the interpretation, by the birds, of what was a head and what a tail. This means that it is not the shape as such that acted as a sign stimulus, since it is the same in both tests, but shape in relation to direction of movement (Lorenz, 1939; Krätzig, 1940).

Although actual proof of the configurational nature of sign stimuli has only been given in relatively few instances, the available evidence points to the conclusion that configurational sign stimuli are the rule rather than an exception. For instance, in each case where 'type-of-movement' acts as a sign stimulus, it is the arrangement of elements in the visual field, in space and in time, that is typical for the releasing situation, the elements themselves being the same in the releasing and the non-releasing situations. Threat posturing in a stickleback (see frontispiece) and threat posturing in a herring gull (Fig. 66) do not add new elements

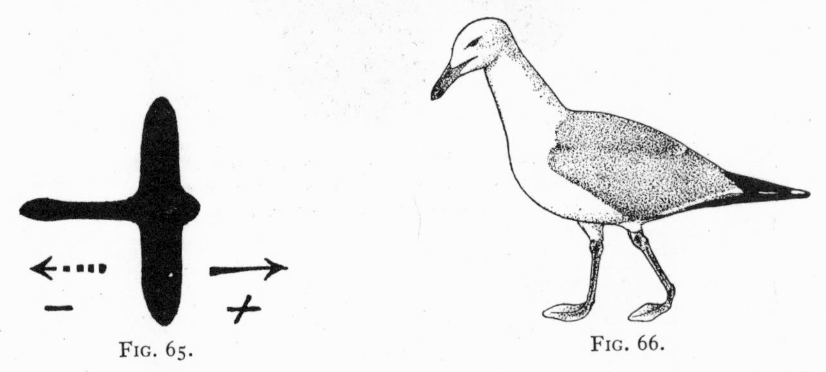

FIG. 65. FIG. 66.

FIG. 65. Model of bird of prey, releasing no escape reactions when sailed to the left, but releasing escape when sailed to the right. After Tinbergen, 1948*a*.

FIG. 66. Threat posture of herring gull. After Tinbergen, 1947*b*.

to the situation; it is only the posture that is changed. Heinroth described (1928) how a young peregrine falcon kept in one room with a number of birds of various species never did them any harm until one happened to fly away from it and promptly released the hunting behaviour.

Thus configurational (*gestalt*) organization of at least visual sign stimuli seems to be the rule. It would be of great value to study this problem in sign stimuli of other sensory qualities too. In the auditory field, especially where, as in birds and Homoptera, melodies and rhythm seem to play a role, in addition to pitch and timbre, configurational stimuli might also be the rule. In the case of chemical and simpler mechanical sign stimuli the situation might be different. In any case, a closer study of this aspect of the sign stimulus problem is highly necessary.

The configurational nature of perception in general has been subject to much discussion, since Köhler (1918) and Wertheimer (1922) called attention to it. The recognition of the *gestalt* principle has done much to open our eyes to the fact that the stimuli to which the animal as a whole responds are not simple measurable entities. However, it should not be forgotten that the *gestalt* principle is not an explanatory principle. The conclusion that a sign stimulus is configurational is merely a provisional way of describing the complexity of the sensory stimulating process. It is thus a challenge rather than a solution, a challenge to analyse the complex system of processes denoted by the convenient collective name 'stimulus'. The *gestalt* conclusion marks one of those points where causal analysis meets a barrier that baffles the experimenter and causes him to revise his methods of study. At such barriers, some of the workers, as it were, capitulate by accepting a mere descriptive term as a causal explanation. This causes a false satisfaction which is a hindrance to further research.

THE STIMULUS SITUATION AS A WHOLE

As was demonstrated above, an IRM usually responds to a complex of two or more sign stimuli. The study of the co-operation of the various sign stimuli releasing a given reaction has led to an important result, which will now be considered.

As we have seen, the level of the motivation determines the degree of sensory stimulation required to release the reaction. When the motivation is relatively strong, suboptimal external stimulation will still bring the total of factors above threshold value. Now, for many reactions the all-or-none law does not hold; they may appear in many different intensities, from a mere vestige of a movement to the complete performance. The lowest intensities of response, being no more than a slight indication of what the animal is tending to do, can only be interpreted by an observer who has watched the whole range of intensities. These incomplete movements are called 'intention' movements (Heinroth, 1911). As a rule, no sharp distinction is possible between intention movements and more complete responses; they form a continuum. The result of this situation is that, with a moderately strong motivation, incomplete external stimulation elicits incomplete reactions. In those cases where the all-or-none law holds, at least for the lower intensities, incomplete external stimulation leads to a decrease, not of the intensity of each individual reaction, but of the number of positive responses to repeated stimulation. This is due to fluctuations in the strength of the motivation, such as can never be completely ruled out. Thus we have two means of observing the effects of incomplete external stimulation, viz. observing (1) the intensity of the reaction, or (2) its frequency, or both.

For instance, the zigzag dance of the male stickleback may consist of a series of two or three separate leaps, or in case of a higher intensity of the motivation, of a much longer series, up to twenty in quick succession. Counting the number of zigzags, therefore, is a means of measuring the reaction's intensity. At the same time, however, the intensity of each separate leap may vary from a relatively slow straight movement to a very quick and sharp turn (Fig. 67).

Now in studies of IRMs it is a most striking fact that deficiencies in sign stimuli always have the same general effect on the reaction, regardless of which part of the stimulus situation is missing or incomplete.

Fig. 67. The zigzag dance of male three-spined stickleback. High intensity (left) and low intensity (right).

The form of the complete reaction is dependent not on what part of the stimulus situation is missing but on how much of it is missing. Although this phenomenon is known to every student of behaviour, it has been studied systematically in only very few cases, and its importance has not been generally realized.

In Chapter II I showed that the sexual flight of *Eumenis* males is released by objects showing the peculiar fluttering type of movement. Apart from this sign stimulus, at least two others are important. First, dark models have a higher releasing value than light ones (Fig. 44); second, distance is a sign stimulus: the closer a model is, the more reactions it releases. Now a white model can release as many reactions as a dark model, provided only that it is close enough. If the brown model is moved 'smoothly' and the white one is 'dancing', the latter again has the same effect as the former, even if the distance is the same in both cases. Deficiency in the sign-stimulus 'darkness' (shade of grey) can be compensated for by either distance or movement. Other experiments proved in the same way that movement can be compensated for by shade or distance.

Seitz (1940) called attention to the same phenomenon in the fish *Astatotilapia strigigena*. As was pointed out by Lorenz (1939), this type of additive co-operation of sign stimuli is quite different from the configurational type of co-operation. Seitz named the phenomenon the *Reizsummenregel* (law of stimulus summation); I will call it the phenomenon of heterogeneous summation.

The fact that the various sign stimuli, though usually qualitatively very different, do not differ in their effect on the motor response as a whole, and can replace each other quantitatively, means that their influence is, somewhere in the central nervous system, added in a purely quantitative way. It is as if the 'impulse flow' started by each sign stimulus, qualitatively different though it may be from the impulses arriving from other receptor fields, acts upon the motor centre in a purely quantitative way. In other words, either before they reach the motor centre or in the motor centre itself, the effects of the external stimuli are brought together and added. As we will see later, there are indications suggesting that this addition is effected in the IRM itself, that is, before the impulses reach the motor centre.

Now, as already stated in Chapter II, the motor part of the response is also highly complex. Even in simpler responses like locomotion of *Labrus*, of the eel, and of other fishes, a whole system of muscles is active in definite co-ordinated patterns. Here again relationships play a role; relationships of sequence and of intensity of contraction. The various integrative principles at work in motor co-ordinations will be discussed later; I shall now merely stress the actual configurational character of most instinctive reactions.

These facts are highly significant in two respects. First, the additive co-operation of sign stimuli releasing a reaction as a whole indicates that the afferent impulses are collected into one single 'container', which acts in a purely quantitative way on the motor centre. Second, the configurational character of the motor response itself shows that the motor centre redispatches the stimuli and distributes them according to configurational principles. This 'container' is, in the terms of neurophysiology, a centre, or a system of centres.

RELEASING AND DIRECTING STIMULI

So far only one aspect of sensory stimulation has been considered; I have been speaking of stimuli as releasing a reaction. However, many sensory stimuli do something else: they direct an activity in relation to the spatial arrangements in the surroundings. The following instances will make the difference clear.

Daphnia, swimming in water with a high carbon dioxide concentration, gather near the surface. The function is obvious: in polluted water the surface layer, being in contact with the air, is relatively rich in oxygen. Analysis shows that two stimuli play a part in this reaction: chemical stimulation by the carbon dioxide, and a visual stimulus. In a glass jar lighted from underneath *Daphnia* will swim downwards as soon as the CO_2 is added. The carbon dioxide merely *releases* the response, which is *directed* by the light.

Young herring gulls react to the parents' alarm calls by taking shelter in the vegetation. While the call releases the reaction it cannot direct it. This was especially evident in the following case. I was watching the family life of gulls from a hide. By making a careless movement I frightened the adult bird, which gave the alarm call and walked away. The chicks promptly reacted to the call by stretching their necks and running for shelter. Because my hiding tent was the nearest shelter, they ran to it, entered it, and crouched at my feet.

Many butterflies react to the colours and to the odours of flowers. In testing the colour vision of *Pieris* and *Vanessa* species, Ilse (1929) found that the butterflies could be stimulated to alight on coloured papers by spraying essential oils into the air. The scent released the response, but the response was directed by the visual stimuli supplied by the papers. The same was found in *Eumenis* in a slightly different way. When kept in cages of fine netting, *Eumenis* could be stimulated to visit coloured papers by hanging a bundle of *Calluna* flowers upwind. Although there was a spatially directed scent current across the cage, the animals did not fly towards the flowers but visited the scentless papers (Tinbergen, Meeuse, Boerema, and Varossieau, 1942). Exactly the same experimental set-up had a different effect on the digger wasp *Philanthus triangulum*. These wasps were not only stimulated but also directed by the scent and flew against the wind as soon as *Calluna* was presented at the windward side, even when the plants were out of sight (Van Beusekom, 1946). Here either the scent or the wind directed the reaction. Therefore, parenthetically, the function of plant odours cannot be judged from the reactions of one species towards them, for different species behave differently.

One and the same stimulus may have both the releasive and the directive effect simultaneously. As was described above, the red patch on the herring gull's mandible releases pecking responses in the chick. When the red patch is situated above the eye not only does it release pecking responses (though fewer than the standard model) but those reactions it does release are directed, in part, to the red patch at its abnormal site (Fig. 68). The fact that some of the reactions are aimed at the bill tip, while proving that the bill tip can direct the response by its shape alone, does not invalidate our conclusion that the red patch has a directing function in addition to its releasing function.

The distinction between releasing and directing stimuli is of considerable importance for our understanding of the physiology of movement. The mechanisms connecting the sensory stimulus with the motor organs are different in the two cases. This difference can be made clear by an analogy. The movements of a steamship are dependent on two mechan-

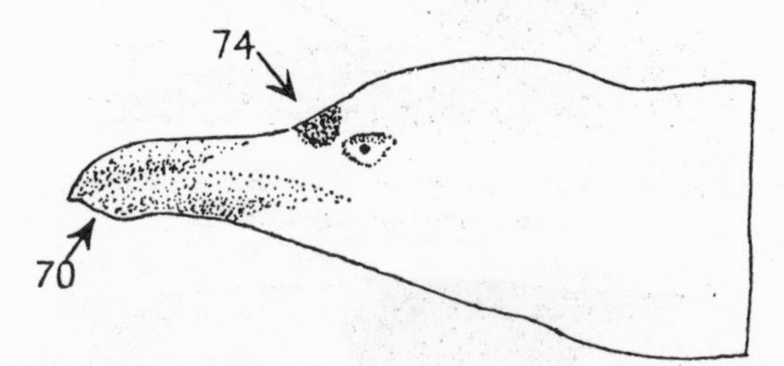

FIG. 68. Directing influence of red patch on herring gull's bill. Numbers indicate numbers of observed reactions of chicks aimed at bill-tip and at red patch, respectively. After Tinbergen, 1948*b*.

isms. The propeller pushes it, the rudder steers it. The forward movement is dependent on a releasing stimulus and can go on without further external stimulation as long as the fuel does not run out. The steering, however, is continuously controlled by new stimuli, coming, eventually, from the external environment. If an outside observer without any understanding of man-made machines studied the ship as we are studying the unknown locomotory machinery of animals he would observe that the releasing stimuli that are responsible for the forward motion of the ship are entirely different from the directing stimuli. The sailing order releasing the departure, the open ocean releasing increase of velocity to full speed, or slowing down the speed, &c., provide stimuli controlling the propeller mechanism. Visual and magnetic stimuli (the latter received by the compass, the ship's magnetic receptor) continuously control the rudder mechanism. It was the realization of the difference between these two types of sensory control that compelled Loeb to consider the problem of oriented movements, for the directing stimuli are those controlling orientation in the widest sense of the word.

The two types of reaction, which I will provisionally name the 'released' reaction and the 'directed' reaction, are often performed simultaneously; as in the ship, the resulting combination gives the impression of being one single reaction. This dual nature of a reaction has been demonstrated in the following cases.

The grey lag goose reacts to eggs that have rolled out of the nest by stretching the neck towards it, bringing the bill behind the egg and with careful balancing movements rolling it back into the nest (Fig. 69). The innate releasing mechanism of this response reacts to relatively few sign stimuli; objects of very different shape and size, provided they have a

Fig. 69. Grey lag goose retrieving egg. After Lorenz and Tinbergen, 1938.

Fig. 70. Grey lag goose attempting to retrieve giant egg. After Lorenz and Tinbergen, 1938.

rounded contour, release it. In spite of the balancing movements the bird sometimes loses control of the egg and then the egg slips sideways. In this case the egg-rolling movement does not always break off, but it may be completed, very much as if it were a vacuum activity. If this happens, the sideways balancing movements are absent. This indicates that the balancing movements are dependent on continuous stimulation from the egg, probably of a tactile nature, while the other component, a movement in the median plane, is not dependent on continuous stimulation but, once released, runs its full course. At the same time, the spatial orientation of this latter movement is, from the very beginning, not influenced by external stimuli. It is an internally fixed pattern. The stereotyped nature of this component was especially obvious when the bird was offered an egg that was much too large. It could not adapt the movement to the abnormal size and got stuck half-way when the egg was pressed between the bill and the breast (Fig. 70). The balancing movements could further be eliminated by presenting the goose with a cylinder. This model, when offered on an entirely flat and smooth surface, made no sideways movements. The result was a straight egg-rolling movement without any sideward deviations of the head (Lorenz and Tinbergen, 1938).

The experimental separation of the directed and the released components of a reaction was accomplished in a different way in the gaping response of nestling thrushes (Tinbergen and Kuenen, 1939). As stated

FIG. 71. Four dummies releasing and directing gaping responses in nestling blackbirds (*Turdus merula*). Arrows indicate part at which gaping is directed. After Tinbergen and Kuenen, 1939.

FIG. 72. Gaping of nestling thrush aimed at the highest of two identical sticks. After Tinbergen and Kuenen, 1939.

above, this reaction is released by mechanical stimuli while the nestlings are still blind, and by visual stimuli during the subsequent period. The mechanical releasing stimulus can be given by jarring the nest slightly in imitation of the alighting parent bird. The direction of the gaping response is not influenced by this stimulus; the nestlings stretched their necks vertically upward, and this is an orientation to gravity. During the visual stage there is a similar difference between releasing and directing stimuli. By means of dummy experiments it was found that the releasing object must have three main characters: it must move, it must be larger than about 3 mm. in diameter, and it must be above the horizontal plane passing through the nestlings' eyes. The response is now no longer directed by gravity, but the young aim at the parent's head. The characteristics of the sign stimuli provided by the head are: it must be an external protrusion of the body outline (Fig. 71), it must be above the body (Fig. 72), it must be closer to the nestlings than the body (Fig. 73), and it must have a diameter of about one-third of that of the body (Fig. 63, p. 77).

The different natures of the two components involved in the gaping response as a whole follows from the difference in the effectors involved: while the released response consists, throughout the nestling period, of stretching the neck and opening the mouth, the muscles involved in keeping the neck vertical (during the first stage) are co-ordinated in a different way when, in the subsequent stage, the neck is pointed to-

FIG. 73. Gaping of nestling thrushes aimed at nearest of two identical sticks. After Tinbergen and Kuenen, 1939.

wards the parent bird's head. In addition, the change in the nature of the effective stimulus situation, the taking over of control by the visual stimuli (which is a process of maturation and not of conditioning), does not take place simultaneously in the two mechanisms. The response to the visual releasing stimuli appears some days in advance of the response to visual directing stimuli; there is an intermediate stage during which the reaction is released by the new visual stimulus but is still steered by gravity. The nestlings present the peculiar spectacle of gaping in response to the sight of the arriving parent, while not pointing their necks towards it (Fig. 74).

In these cases, therefore, it is obvious that the reaction considered, though giving the impression of being an individual unit of behaviour, is a combination of two simultaneously linked components.

Now it is very probable that the same holds true for a great number of oriented movements—for instance in many cases of oriented locomotion. One component is a more or less fixed pattern which is controlled by external releasing stimuli in co-operation with internal motivating factors and which, once released, is integrated by internal mechanisms only, often quite independently of further external stimulation. The second component is not a fixed, internally co-ordinated pattern, but is a sequence of reactions to external stimuli that continuously correct the direction of the movement in relation to the spatial properties of the environment. Lorenz, who first made this distinction (1937*a*), pointed out that the oriented movements of animals called tropisms by Loeb and taxes by Kühn are thus integrations of these two components. He called the former component *Erbkoordination*, for which I propose the

FIG. 74. Nestling thrushes gaping in response to visual stimulus but not directed by it. After Tinbergen and Kuenen, 1939.

English term 'fixed pattern'. The orientation mechanism was called the *Taxiskomponente*, using Kühn's term but restricting its meaning. We will accept Lorenz's distinction between the oriented movement as a whole and this restricted orientation or correction mechanism which is its component, and call the latter a 'taxis'.

The reasons for splitting the older concept appear more urgent when it is realized that there are cases in which taxis and fixed pattern are not linked simultaneously but are performed in succession. A clear instance of this is the food-catching behaviour of a frog. When a frog sees a fly it makes a sideways turn which results in its facing the fly fair and square. If the fly then comes within shooting distance the frog will flip out its tongue (Fig. 75). The first movement is a taxis; it can be a mere turning towards the prey without any forward motion and in that event is a pure taxis. If, as is often the case, the prey keeps moving about, the taxis will be combined with locomotory patterns and a more or less complex oriented movement will be the result. The actual tongue-flipping, after having been released by the adequate visual stimulus situation, is not corrected. If the prey moves in time, the frog will miss it. The tongue movement is a true fixed pattern, though one of the more simple types.

To sum up: an oriented movement as a whole may be either a succes-

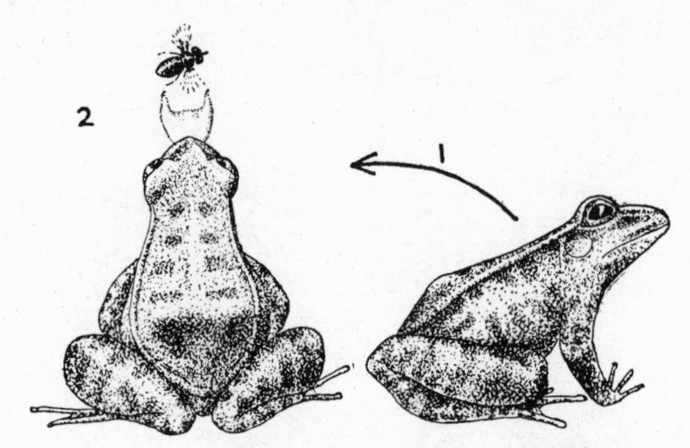

Fig. 75. Aiming and subsequent shooting in frog.

Fig. 76. Aiming and subsequent shooting in primitive man.

sive or a simultaneous integration of two different types of patterns: (1) a taxis, being an integration of reflex-like reactions, responsible for the orientation; and (2) a fixed pattern, being an integration of muscle contractions controlled from within. Both taxis and fixed pattern may be of highly varying degrees of complexity. The further analysis of fixed patterns will be undertaken in Chapter V. I shall now proceed to discuss the various types of taxes or orientation movements.

TYPES OF ORIENTATION

The hitherto confused interpretation of orientation movements is one more example of the unhappy effects of premature generalization based on observations of a restricted number of animal species. Loeb generalized his important concept of tropisms by assuming that every oriented movement must be a tropism. Jennings pointed out that certain protozoans behaved on the basis of trial and error, which, under certain conditions, could result in something very similar to spatially directed behaviour. Although Jennings himself was much more aware of the multiformity of physiological mechanisms than Loeb, there followed a controversy between supporters of the trial-and-error principle

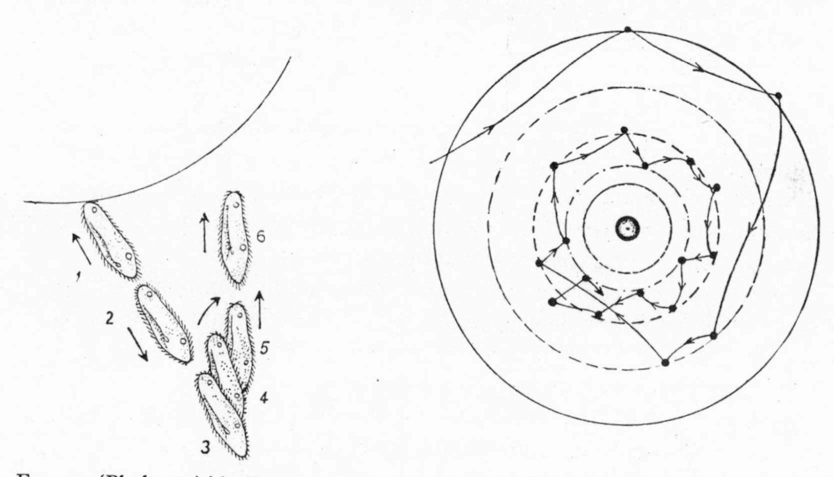

FIG. 77. 'Phobotaxis' in *Paramaecium* in vicinity of CO_2-bubble. After Jennings, 1906, from Kühn, 1919.

and the tropism principle which lasted until it was realized that both mechanisms existed.

Kühn (1919) has given a tentative classification of the orientation movements. Although he did not split the oriented movement into its two components, his classification can for the most part be applied to the isolated taxis component. He distinguishes trial-and-error orientation from real directed movement or 'topotaxis'.

The classic example of trial and error, or phobotaxis, as Kühn named it, is the movement of *Paramaecium* in the neighbourhood of a bubble of carbon dioxide (Fig. 77). The movements are random. Whenever a *Paramaecium* arrives at a region where the concentration is too high, it swims backwards, turns through a certain angle, and proceeds again. The new direction is not related to the source of stimulation. This same phenomenon of withdrawal and subsequent movement in some other direction occurs every time the *Paramaecium* moves into a high concentration of carbon dioxide. The result is that the animal remains in a region of optimal concentration, where it is captured in a 'physiological trap'. Although the result is a definite spatial distribution of the animals, the correction movements are not directed in relation to environment, and they offer no orientation problem; they are not taxes.

The genuine oriented movements enabling an animal to follow a path directed by environmental stimuli have been treated extensively by Fraenkel and Gunn (1940). These authors showed that we have to distinguish even more types of orienting movements than did Kühn. Although I will follow their classification in its main outlines, I think the problem may be clarified by applying our knowledge of the dual nature of the oriented movement.

Klinotaxis

The most simple orientation movement is the klinotaxis. The classical example of this type is the maggot of *Calliphora erythrocephala*, the blow-fly or bluebottle, which, for some days before pupation, has a tendency to move away from the light. In crawling it turns its head from time to time alternately to the left and to the right. It is the differences in intensity between the successive light stimuli impinging upon the head during these sideward movements that determine the animal's course. Although the light receptors are not known, it is certain that they must be located at the front of the head, and as long as the maggot crawls away from the light its body shades the receptors (Fig. 78). Gunn (Fraenkel and Gunn, 1940) devised an ingenious experiment which proves that the receptor as such is incapable of localization of direction. A maggot was kept in the dim light provided by a lamp hanging centrally over a sheet of wet ground glass. A second light, also hanging centrally over the experimental field, can be switched on for a moment. If this light is switched on for an instant each time the animal is moving its head to the left, it can be forced to move in a circular course to the right, and, of course, vice versa. The animal gives the same response when the light is all from above, or uniform, or even when it is presented as a horizontal beam. Thus it was proved that the orientation was independent of the actual direction of the light, but dependent on the different intensities of two stimuli administered in succession.

Klinotaxis, though relatively rare in the visual sensory field, where more specialized receptors capable of true direction-finding have evolved in many animals, seems to be found especially in orientation to chemical stimuli. The reactions of *Planaria* to a chemically stimulating bait (freshly cut pieces of aquatic snails) is an example. At distances from about 8 cm. to 2 cm., in a region with a moderately steep gradient of concentration, a straight course towards the bait is dependent on side-to-side movements of the head (Koehler, 1932; Fraenkel in Fraenkel and Gunn, 1940).

Tropotaxis

Many animals are able to direct their movements by a simultaneous comparison of intensities of the stimuli acting upon two receptors or upon two parts of one receptor. In the experiment with maggots described above, both left and right receptors receive the same amount

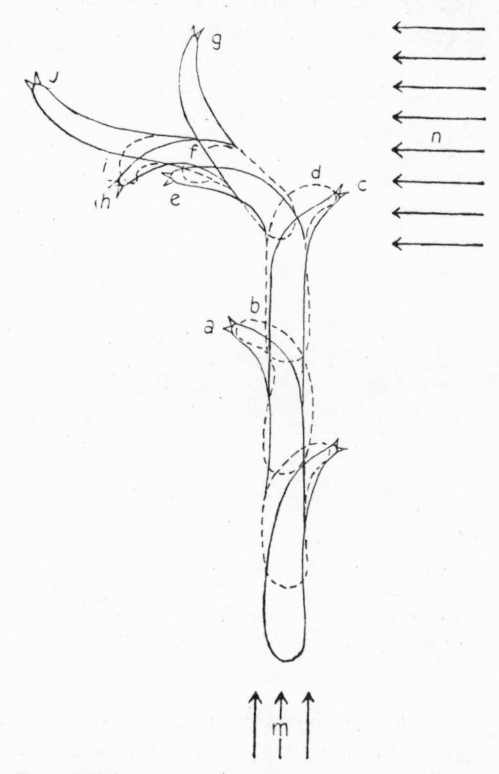

FIG. 78. Klinotaxis in blowfly maggot. The maggot expands to *a*, contracts to *b*, expands to *c*, &c., with the light *m* on. At *d*, *m* is put out and *n* put on. The lateral movement to *e* is somewhat larger than *a*; the next one (*g*) increases the light falling on the receptors and is corrected by a swing over to *h*. After Fraenkel and Gunn, 1940.

of light. Yet the animal turns away from the left if the flash occurred during a turn to the left. Simultaneous comparison of the intensities on the left and right sides is the basis of a type of taxis called 'tropotaxis'. Unequal stimulation of the two receptors causes the animal to move in such a way as to restore balance of stimulation, viz. to turn towards the more strongly stimulated side in a positive taxis, to turn to the other side in a negative (avoiding) taxis. This mechanism works especially well in animals with eyes capable of localization of direction, so that a very slight deviation will cause inequalities of stimulation. Eye cups of the planarian type, and especially compound eyes of the Arthropod type serve this purpose very well. An indication of tropotaxis is the circular movements of animals from which the receptor of one side has been removed. The constant inequality of stimulation causes continuous turning. Fig. 79 shows the tracks of the pill-bug *Armadillidium* after

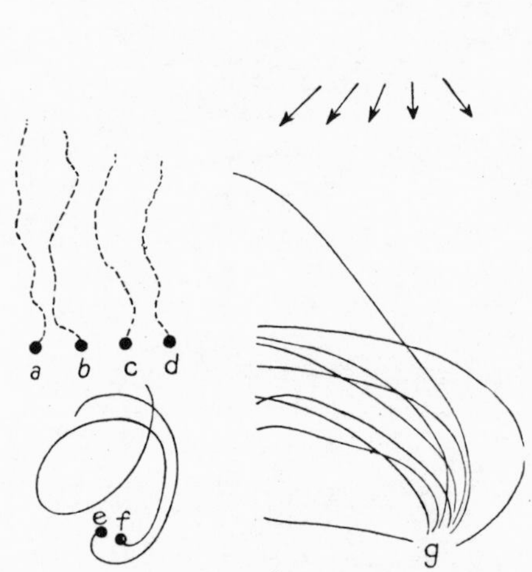

FIG. 79. Tracks of photo-positive *Armadillidium* blinded on the right side. *a–d* in darkness; *e*, *f* with the light overhead; *g* in a beam, the direction of which is indicated by the arrows. After K. Henke, 1930.

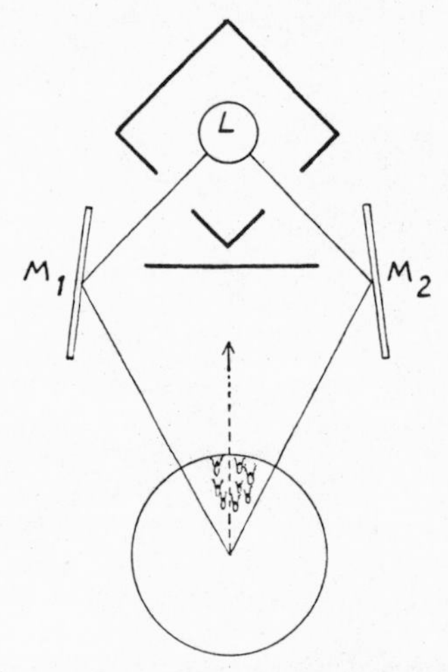

FIG. 80. Two-light experiment with *Daphnia*. *L*, light: M_1 and M_2, mirrors. After Kühn, 1919.

blinding one eye. The turning is directed towards the intact side, the experiments being made with positively phototactic animals. A second type of test is the ingenious two-stimulus experiment. If, for instance, two lights are place in front of a positively phototropotactic animal, it will not move towards either of the two lights but it will go between them. An example given by Kühn is the positive phototaxis shown by *Daphnia* after stimulation by carbon dioxide (Fig. 80).

This is the type of oriented movement called tropism by Loeb. The exact mechanism is not known in any particular case; it seems certain, however, that it is less simple than originally supposed.

Koehler (1932) has shown that chemical stimuli may act in this way on *Planaria* in a very steep gradient of concentration, viz. at a distance of less than about 2 cm. from the bait used in his experiments. In one experiment a double chemical stimulus was given by a bent tube filled with snail's blood, the two open ends being presented at exactly equal distances from the two chemoreceptors. When this two-source stimulus is very slowly moved away from the animal and care is taken to keep the distance between tube and receptor equal on both sides, the animal will move in a straight course directed towards their midpoint (Fig. 81).

'Dorsal Light Reaction'

A reaction very similar to tropotaxis is the dorsal (or ventral) light reaction found in many aquatic animals and enabling them to keep the dorsal (or ventral) side uppermost. In many cases it seems to be based on a balanced stimulation of two eyes. For instance, the crustacean *Lepidiurus apus* swims with the dorsal side uppermost. If the direction of the light is changed, reorientation is achieved by sideward rolling (Seifert, 1930). If one of the eyes is painted over, the animal shows continuous rolling and circling movements towards the intact side. The brine shrimp *Artemia salina* has a ventral light reaction. When the light is moved from the top to below, *Artemia* attains reorientation not by rolling sideways but by somersaulting. If one eye is removed, it shows continuous sideward rolling.

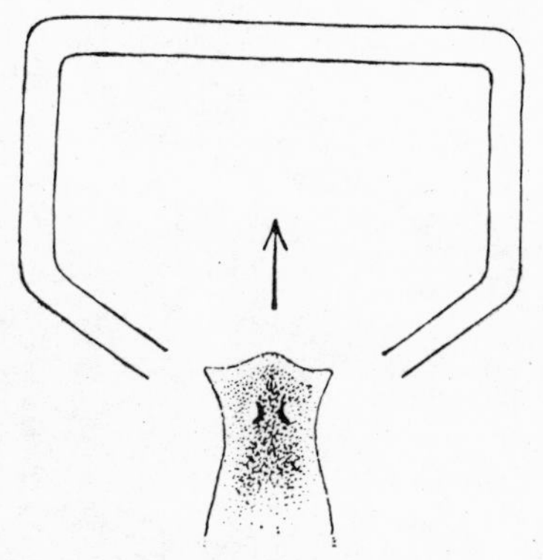

FIG. 81. Two-stimulus experiment demonstrating chemotropotaxis in *Planaria*. After Koehler, 1932.

Fraenkel and Gunn distinguish this type of orientation movement sharply from both telo- and tropotaxis and from the light compass reaction, they state:

> This reaction is similar to tropotaxis and telo-taxis in that the animal places itself symmetrically with respect to a single incident light, but it is unlike these two taxes in that the animal does not move towards the light but—if it moves at all—at right angles to it. On the other hand, in the light compass reaction the animal does not place itself symmetrically but it does, as in the dorsal light reaction, move at an angle to the light rays (pp. 120–1).

I believe that Fraenkel and Gunn are here confusing oriented movement as a whole (taxis-fixed pattern combination) with the pure taxis component. The mechanism studied in taxes is the sensory-neuromuscular mechanism involved in *acquiring* the oriented position, not the locomotory rhythm shown in addition. That the type of locomotion is irrelevant for the orientation problem is clear from the authors' own remark quoted above, that the animal need not move, that is to say, that locomotion may be absent. Rolling may or may not be accompanied by locomotion; and the rolling is the orientation movement. The problem is therefore whether we have to do with a special case of tropotaxis or of telotaxis.

The two-light experiments and those with unilaterally blinded animals mentioned above, making use of *Lepidurus* and *Artemia*, clearly show that balance of stimulation and, therefore, tropotaxis is involved. Less

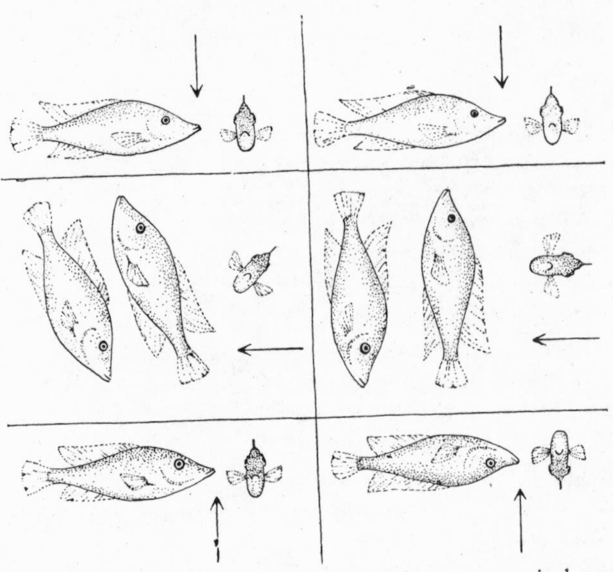

FIG. 82. Dorsal light reaction in *Crenilabrus rostratus* in beams coming from different directions. Left: intact; right: without labyrinth. After von Holst, 1935*a*.

simple is the situation in the fish *Crenilabrus* studied by von Holst (1935*a*). The normal position in this fish, as in many others, is maintained by both gravity receptors and the eyes. If the static receptors are removed, a pure dorsal light reaction is observed. Unilaterally blinded fish at first roll to the intact side (Fig. 82). After some time, however, rolling ceases and normal visual orientation is maintained with one eye. Thus a telotaxis appears to take over.

Telotaxis

In many animals with well-developed eyes another mechanism may make possible the pursuit of a straight path. This type of taxis does not depend on a balance of stimulation between two receptors. When one eye of a dragonfly or a dragonfly larva is blinded, it can still move straight towards and shoot its mask at a prey. This taxis is called 'telotaxis'. It seems to be dependent on a fixation of the stimulus source.

In the two-stimulus test this taxis leads the animal to one of the two sources (Fig. 83). Telotaxis is often involved in reactions to definite objects such as prey. The object may give a relatively simple stimulus such as 'a light', but more often it gives a configurational stimulus. For instance, the sexual pursuit of the male grayling (p. 40) is telotactically oriented towards a passing female. The stimuli enabling nestling thrushes to aim their gaping response at the head of the parent bird (p. 76) are of a complex, configurational kind; they release a true telotaxis. Contrary to the opinion of Fraenkel and Gunn, I consider the last phase of

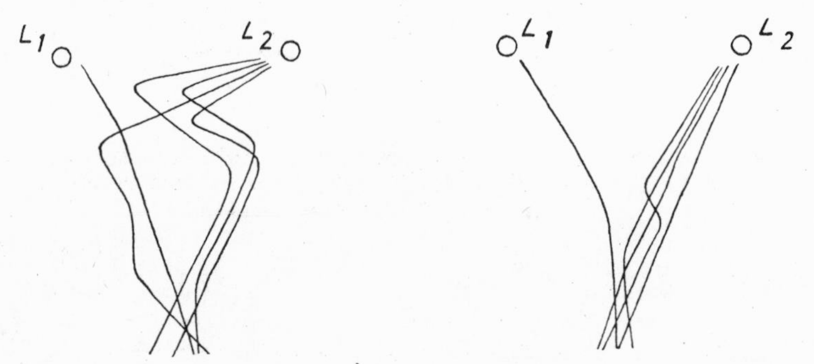

FIG. 83. Tracks of hermit crabs in a two-light experiment. L_1 and L_2, lights. After von Buddenbrock, 1922.

the oriented movements of a fossorial wasp returning to its nesting hole a telotaxis, directed towards one definite spot in a configurational field of landmarks, a type of taxis which I previously called 'pharotaxis' (Tinbergen and Kruyt, 1938; Tinbergen, 1942). The main difference from other types of complex telotaxis is that the stimulus situation has acquired its influence by conditioning.

Menotaxis

A special case of telotaxis is the menotaxis or light compass reaction. Menotaxis resembles telotaxis in that it is dependent on one eye only. However, the course followed is directed neither straight towards an object nor straight away from it. The angle between the direction of locomotion (or rather of the body axis) and the direction of the stimulus source is kept constant. The source of stimulation in the classical examples of light compass orientation is the sun. Fig. 84 illustrates one example observed in ants by Santschi (1911). The influence of the sun was demonstrated by the use of a mirror.

Brun (1914) kept a captured homing ant under a lightproof box and released it after 2½ hours, in which period the sun had travelled 37° in horizontal projection. The ant changed its direction accordingly (Fig. 85). Similar observations have been made by Wolf (1926, 1928) in bees. Hive bees, though normally using landmarks, turn to menotaxis when flying over quite bare stretches of country offering no adequate landmarks.

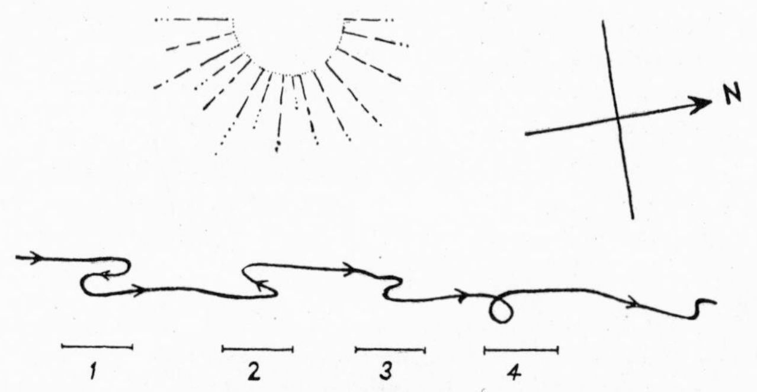

FIG. 84. Mirror experiment with homing ant. At 1, 2, 3, and 4 the light from the sun was intercepted and the image of the sun was projected from the opposite side by means of a mirror. After Santschi, 1911.

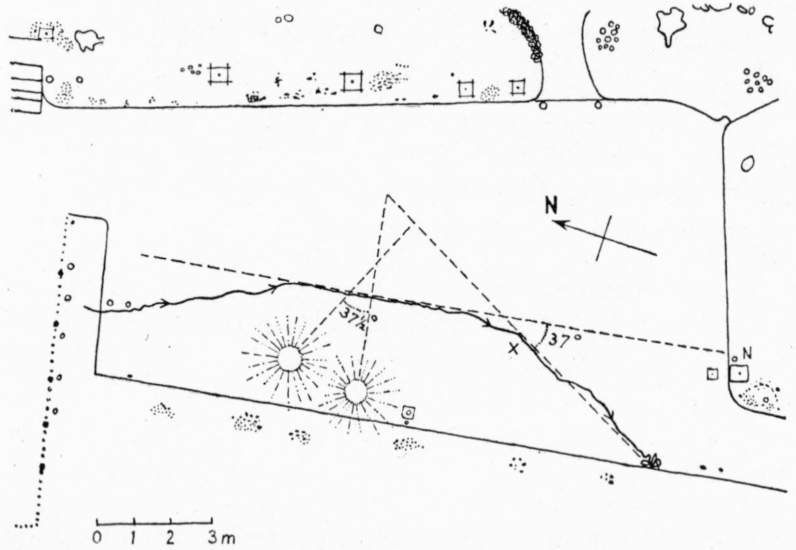

FIG. 85. Demonstration of menotaxis in homing ant. Cross indicates spot where the ant was kept in the dark during $2\frac{1}{2}$ hours. After Brun, 1914.

The light compass reaction offers essentially the same problem as telotaxis. Common to both taxes is the fact that the animal's position is kept constant in relation to the source of stimulation. It does not seem to be essential whether the image has to be fixed (as it is thought to be in telotaxis) or whether it has to be kept at any other place on the retina (as supposed in the case of light compass orientation).

In menotaxis, too, conditioning processes usually play a part.

'Mnemotaxis'

In Kühn's original classification another taxis was distinguished by the name 'mnemotaxis'. This name was given to the complicated type of orientation shown, for example, by homing wasps and bees. It was considered a separate type for two reasons. First, the orientation is not dependent on one single object but on a number of landmarks. Second, memory is involved; the homing animal is conditioned to the landmarks. Now as I have argued above, conditioning as such is no criterion for a classification of orientation mechanisms. It concerns the developmental history of the taxis, not its mechanisms.

The other argument, that of the complexity of the situation, must be considered more closely. Fraenkel (1931) has claimed that mnemotaxis is nothing but a sequence of taxes of other types. This, however, is an assumption lacking proof. Recent studies on the homing of fossorial wasps have shown that, although the homing flight of the strong fliers among them is a chain of different orientation activities, the last stage cannot be considered a mere chain of simple taxes. Van Beusekom (1948), for instance, showed that *Philanthus triangulum* uses a number of landmarks simultaneously. One of his experiments is illustrated in Fig. 86. The training situation consisted of a square block close to the nest and a 'tree' 1 metre from the nest. In the test, the square was moved to the right and turned 45°. The wasp chose corner 1 when the tree was moved to position 1, and she chose corner 2 when the tree was opposite this corner. This suggests that the wasp uses both block and tree up to the last moment.

Other tests showed the simultaneous use of landmarks still more clearly. When the wasp was trained to use a circle of pine-cones around the nest, and some of the cones were moved to the right and some to the left, or some were taken away and the remaining cones moved a little distance, the wasp's choices showed that it did not react to any particular part of the total situation but to a combination of them all.

Fig. 87 illustrates a number of tests in a natural situation where a twig and three pine-cones were used as beacons. When these beacons were moved separately or in partial combinations, it became clear that the wasp did not orient itself to one particular beacon but to the whole constellation (Tinbergen and Kruyt, 1938). When the course followed by the wasp was observed (Figs. 87 and 88), another important fact became evident: orientation was not dependent on the course followed, for the flight was highly variable. In other words, orientation is not dependent on stimulation of special parts of the retina by images of special landmarks. The only constant feature was that the wasp always flew towards the nest entrance. In other words, it aims at a spot

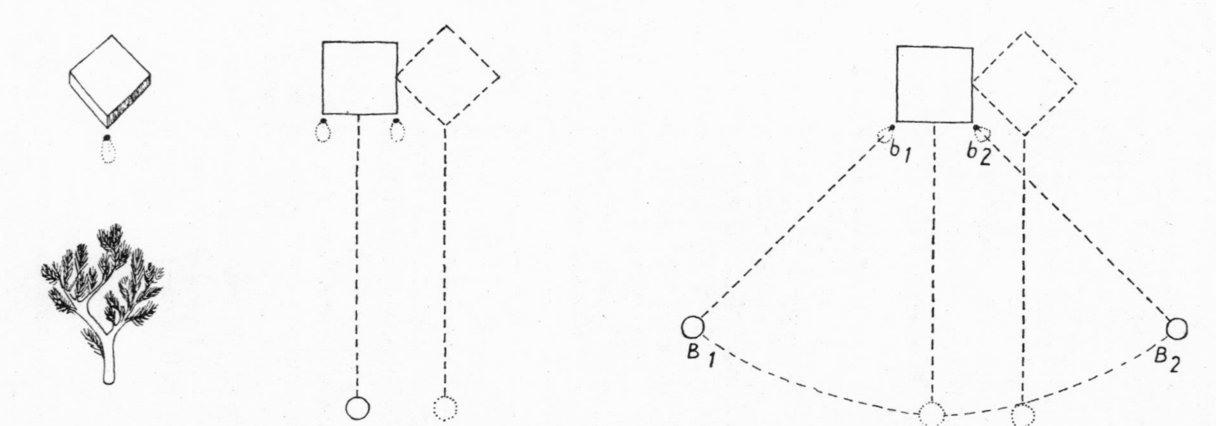

FIG. 86. Experiments demonstrating use of two landmarks by homing digger wasp, *Philanthus triangulum*. Left: training situation. Centre: first test, in which both block and tree were displaced and block turned 45°. Right: block displaced and turned 45°, tree alternately in B_1 and B_2. Wasp alights at b_1 when tree is at B_1 and at b_2 when tree is at B_2. After Van Beusekom, 1948.

determined by its spatial relationships with the whole configuration of land-marks. Now when the cases of telotaxis mentioned above are compared with this type of orientation, it is obvious that the two types differ only

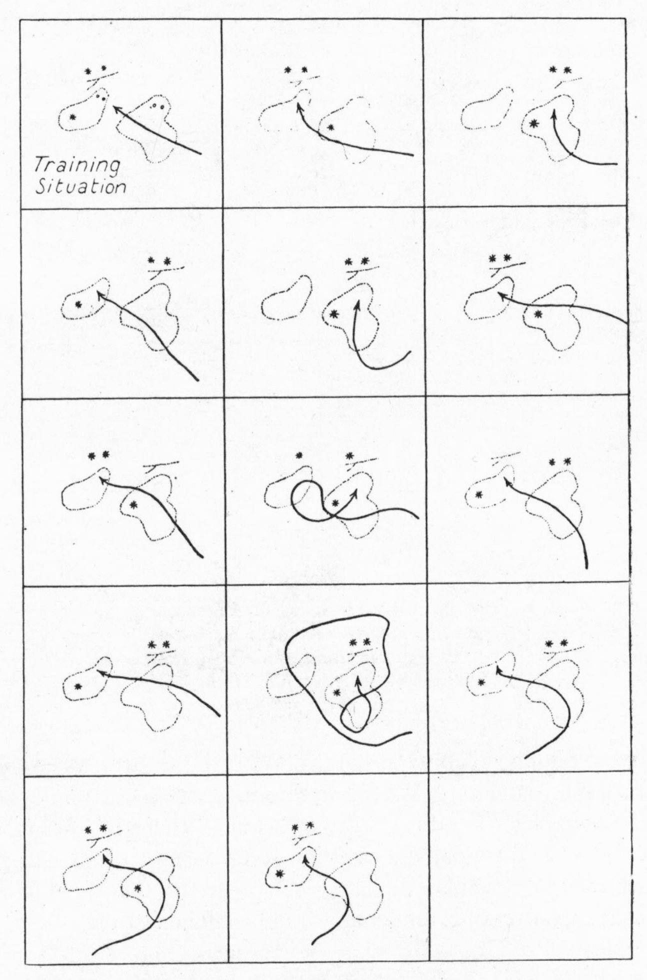

FIG. 87. Homing flights of *Philanthus triangulum* in displacement tests with three pine-cones, indicated by stars, and a forked twig. After Tinbergen and Kruyt, 1938.

in degree. The two differences are: (1) the configuration is, in the case of the homing wasp, more complicated than in the case of, say, a dragon-fly capturing a mosquito or of a nestling thrush gaping towards its parent's head, and (2) the wasp probably does not fix the nest entrance, whereas simple telotaxis is claimed to be preceded by fixation movements.

Whether this difference is essential remains to be seen. It would seem, therefore, that there are only differences of degree between telotaxis, menotaxis, and the type considered here, or pharotaxis. The

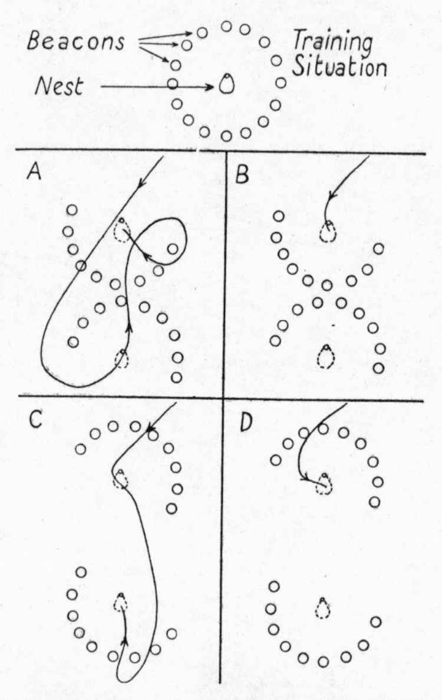

FIG. 88. Homing flights of *Philanthus triangulum* in displacement tests with parts of the training situation (circle of pine-cones). After Tinbergen and Kruyt, 1938.

simplicity of telotaxis is only apparent. With these three taxes there is a configurational visual field. Within this configuration one point is singled out. Locomotion is then directed in relation to that goal. Whether it is straight towards the goal or whether it takes any other direction is irrelevant to the problem of the mechanism of taxes, which is the mechanism of correcting movements, not of locomotion.

V

AN ATTEMPT AT A SYNTHESIS

RECAPITULATION

WE have now arrived at a point where it is necessary to review our results in order to evaluate and appreciate their significance in relation to our main problem, that is, the problem of the causation of instinctive behaviour.

The foregoing chapters have led to the following conclusions.

Instinctive behaviour is dependent on external and internal causal factors. The external factors, or sensory stimuli, are of a much simpler nature than our knowledge of the potential capacities of the sense organs would make us expect. Yet they are not so simple as the word 'stimulus' would suggest, for the 'sign stimuli' have *gestalt* character, that is to say, they release configurational receptive processes. The various sign stimuli required for the release of an instinctive activity co-operate according to the rule of heterogeneous summation. These facts led us to the postulation of Innate Releasing Mechanisms, one of which is possessed by each separate reaction. Apart from releasing stimuli, directing stimuli play a part, enabling or forcing the animal to orient itself in relation to the environment. The internal causal factors controlling, qualitatively and quantitatively, the motivation of the animal may be of three kinds: hormones, internal sensory stimuli, and, perhaps, intrinsic or automatic nervous impulses generated by the central nervous system itself. Instinctive 'reactions' are of varying degrees of complexity; even the simplest type, the 'fixed pattern', depends on a system of muscle contractions which is of a configurational character.

These results are incomplete in more than one respect. First, the evidence is still very fragmentary, and the generalizations are still of a very tentative nature. Second, the work done thus far has been mainly analytical, and no attempt has yet been made to combine the separate conclusions into a picture of the causal structure underlying instinctive behaviour as a whole. We have, however, gained one thing: we are realizing more and more clearly that the physiological mechanisms underlying instinctive behaviour are much more complicated than we were able to see at the start. Previous attempts at synthesis, such as Pavlov's reflex theory and Loeb's tropism theory, now appear to be grotesque simplifications.

While thus realizing both the relative paucity of analytical data and the complexity of the causal structure, we will nevertheless venture to sketch, in rough outline, a synthetic picture of the organization of the partial problems within the main problem as a whole.

DIFFERENCES IN DEGREE OF COMPLEXITY OF 'REACTIONS'

So far I have been using the terms 'reaction', 'motor response', 'behaviour pattern', 'movement' for muscle contractions of very different degrees of complexity. This fact is of paramount importance, and I will emphasize it by presenting some more instances.

As we have seen, the swimming of an eel is a relatively simple movement. In every somite there is alternating contraction of the longitudinal muscles of the right and the left half of the trunk. In addition, the pendulum movements of successive somites are slightly out of step, each somite contracting a short time after its predecessor. The result is the propagation of the well-known sinusoid contraction waves along the body axis (Gray, 1936).

The swimming movements of a fish like *Labrus* or *Sargus*, as described by von Holst (1935*b*, 1937), are more complex. The pectoral fins, moving back and forth in alternation, are also in step with the dorsal, caudal, and anal fins, each of which makes pendulum movements as well.

The movement of a male stickleback ventilating its eggs is of a similar type. The pectorals make pendulum movements alternately. This motion is directed forward, resulting in a water current from the fish to the nest. In order to counteract the backward push this exerts upon the fish, forward swimming movements of the tail are made in absolute synchronization with the rhythm of the pectorals.

Although locomotion might be considered merely an element of a 'reaction' in the sense in which I have been using this term, the stickleback's ventilating movement is a complete reaction, responding in part to a chemical stimulus emanating from the nest.

The reaction of a gallinaceous chick to a flying bird of prey is, again, somewhat more complicated. It may consist of merely crouching, but often it consists of running to shelter provided by the mother or by vegetation, crouching, and continuously watching the predator's movements.

Finally, a male stickleback in reproductive condition responds to visual and temperature stimuli of a rather simple type by behaviour of a very complicated pattern: it settles on a territory, fights other males, starts to build a nest, courts females, and so on.

HIERARCHICAL ORGANIZATION

A closer study of these differences in complexity leads us to the conclusion that the mechanisms underlying these reactions are arranged in a hierarchical system, in which we must distinguish between various levels of integration.

The reproductive behaviour of the male stickleback may be taken as an example.

In spring, the gradual increase in length of day brings the males into a condition of increased reproductive motivation, which drives them to migrate into shallow fresh water. Here, as we have seen, a rise in temperature, together with a visual stimulus situation received from a suitable territory, releases the reproductive pattern as a whole. The male settles on the territory, its erythrophores expand, it reacts to strangers by fighting, and starts to build a nest. Now, whereas both nest-building and fighting depend on activation of the reproductive drive as a whole, no observer can predict which one of the two patterns will be shown at any given moment. Fighting, for instance, has to be released by a specific stimulus, viz. 'red male intruding into the territory'. Building is not released by this stimulus situation but depends on other stimuli. Thus these two activities, though both depend on activation of the reproductive drive as a whole, are also dependent on additional (external) factors. The influence of these latter factors is, however, restricted; they act upon either fighting or building, not on the reproductive drive as a whole.

Now the stimulus situation 'red male intruding', while releasing the fighting drive, does not determine which one of the five types of fighting will be shown. This is determined by additional, still more specific stimuli. For instance, when the stranger bites, the owner of the territory will bite in return; when the stranger threatens, the owner will threaten back; when the stranger flees, the owner will chase it; and so on.

Thus the effect of a stimulus situation on the animal may be of different kinds. The visual stimulus 'suitable territory' activates both fighting and nest-building; the visual situation 'red male in territory' is specific in releasing fighting, but it merely causes a general readiness to fight and does not determine the type of fighting. Which one of the five motor responses belonging to the fighting pattern will be shown depends on sign stimuli that are still more restricted in effect. The tactile stimulus 'male biting' releases one type of fighting, the visual stimulus 'male threatening' another type. The stimulus situations are not of an essentially different order in all these cases, but the results are. They belong to different levels of integration and, moreover, they are organized in a hierarchical system, like the staff organization of an army or of other human organizations. Fig. 89 illustrates the principle. The facts (1) that at each of the levels an external stimulus can have a specific releasing influence and (2) that each reaction has its own motor pattern, mean that there is a hierarchical system of IRMs and of motor centres.

So far as we can judge at present, each IRM is able to collect sensory impulses according to the rule of heterogeneous summation, and each motor centre controls a configurational pattern of muscle contractions.

The principle of hierarchical organization has been tested in but three cases: the digger wasp *Ammophila campestris* (Baerends, 1941), the three-spined stickleback (Tinbergen, 1942), and the turkey (Räber, 1948),

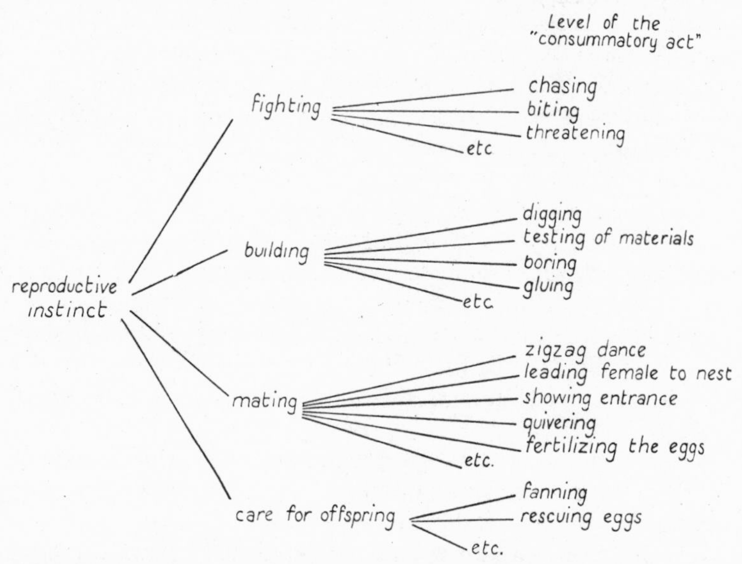

FIG. 89. The principle of hierarchical organization illustrated by the reproductive instinct of the male three-spined stickleback. After Tinbergen, 1942.

and although the principle is undoubtedly sound, nearly nothing is known in detail about the way it works out in the various drives and in different species of animals. Before a more detailed discussion can be attempted, a closer consideration of motor responses is necessary.

APPETITIVE BEHAVIOUR AND CONSUMMATORY ACT

The activation of a centre of the lowest level usually, perhaps always, reaults in a relatively simple motor response: biting, chasing, threatening, &c., in the case of fighting in the stickleback; actual eating, actual escape, actual coition, &c., in other instincts. This type of response has been the object of our analysis in most of the cases treated in the preceding chapters. This is no accident; it is the natural outcome of the tendency to analyse which leads to a conscious or (more often) unconscious selection of relatively simple and stereotyped phenomena.

These relatively simple responses are, usually, the end of a bout of prolonged activity, and their performance seems to 'satisfy' the animal, that is to say, to bring about a sudden drop of motivation. This means that such an end-response consumes the specific impulses responsible for its activation. Fighting, eating, mating, 'playing the broken wing', &c., are, as a rule, 'self-exhausting'. Craig (1918), in a most remarkable paper that has not received the attention it deserves, was the first to single out these elements of behaviour; he called them 'consummatory actions'. Lorenz (1937*b*), realizing that they constitute the most characteristic components of instinctive behaviour, that is to say those components that can be most easily recognized by the form of the movement, called them *Instinkthandlungen*, thereby greatly narrowing the concept of instinctive act. This use of the term gives rise to continuous misunderstandings and hence should be dropped.

The centres of this lower type of movement rarely respond to the external stimulus situation alone. As a rule, they get their internal impulses from a superordinated centre. The activation of these higher centres may result either in a mere increase in readiness of the animal to react with one of a number of consummatory actions, or, more often, in a type of movement often called 'random movement', 'exploratory behaviour', 'seeking behaviour', or the like. Contrary to the consummatory action it is not characterized by a stereotyped motor pattern, but rather by (1) its variability and plasticity and (2) its purposiveness. The animal in which a major drive, like the hunting drive, the nest-building drive, the mating drive, is activated starts searching or exploratory excursions which last until a situation is found which provides the animal with the stimuli adequate for releasing the consummatory act.

As mentioned above, Craig recognized these two types of behaviour, viz. the variable striving behaviour and the rigid consummatory action; and, moreover, he saw their mutual relationships as components of instinctive behaviour as a whole. He called the introductory striving or searching phase 'appetitive behaviour' to stress the fact that the animal is striving to attain some end.

Appetitive behaviour may be a very simple introduction to a consummatory action, as in the case of a frog catching a prey; the preparatory taxis (turning towards the prey) is true purposive behaviour, and is continued or repeated until the prey is within range and in the median plane.

More complicated is the appetitive phase of feeding in a *Planaria* mounting a stream against a scent-loaded current.

Heinroth (1910) describes a still higher form of appetitive behaviour in mated ducks exploring the country for a nesting-hole.

In extreme cases the appetitive behaviour may be prolonged and highly adaptable, as in the migratory behaviour of animals.

It will be clear, therefore, that this distinction between appetitive behaviour and consummatory act separates the behaviour as a whole into two components of entirely different character. The consummatory act is relatively simple; at its most complex, it is a chain of reactions, each of which may be a simultaneous combination of a taxis and a fixed pattern. But appetitive behaviour is a true purposive activity, offering all the problems of plasticity, adaptiveness, and of complex integration that baffle the scientist in his study of behaviour as a whole. Appetitive behaviour is a conglomerate of many elements of very different order, of reflexes, of simple patterns like locomotion, of conditioned reactions, of 'insight' behaviour, and so on. As a result it is a true challenge to objective science, and therefore the discrimination between appetitive behaviour and consummatory act is but a first step of our analysis.

A consideration of the relationships between appetitive behaviour and consummatory act is important for our understanding of the nature of striving in animals. It is often stressed that animals are striving towards the attainment of a certain end or goal. Lorenz has pointed out not only that purposiveness, the striving towards an end, is typical only of appetitive behaviour and not of consummatory actions, but also that the end of purposive behaviour is not the attainment of an object or a situation itself, but the performance of the consummatory action, which is attained as a consequence of the animal's arrival at an external situation which provides the special sign stimuli releasing the consummatory act. Even psychologists who have watched hundreds of rats running a maze rarely realize that, strictly speaking, it is not the litter or the food the animal is striving towards, but the performance itself of the maternal activities or eating.

Holzapfel (1940) has shown that there is one apparent exception to this rule: appetitive behaviour may also lead to rest or sleep. As I hope to show further below, this exception is only apparent, because rest and sleep are true consummatory actions, dependent on activation of a centre exactly as with other consummatory actions.

Whereas the consummatory act seems to be dependent on the centres of the lowest level of instinctive behaviour, appetitive behaviour may be activated by centres of all the levels above that of the consummatory act. As has been pointed out by Baerends (1941), appetitive behaviour by no means always leads directly to the performance of a consummatory act. For instance, the hunting of a peregrine falcon usually begins with relatively random roaming around its hunting territory, visiting and exploring many different places miles apart. This first phase of appetitive behaviour may lead to different ways of catching prey, each dependent on special stimulation by a potential prey. It is continued until such a special stimulus situation is found: a flock of teal executing flight manoeuvres, a sick gull swimming apart from the flock, or even a running mouse. Each of these situations may cause the falcon to abandon its 'random' searching. But what follows then is not yet a consummatory action, but appetitive behaviour of a new, more specialized and more restricted kind. The flock of teal releases a series of sham attacks serving to isolate one or a few individuals from the main body of the flock. Only after this is achieved is the final swoop released, followed by capturing, killing, plucking, and eating, which is a relatively simple and stereotyped chain of consummatory acts. The sick gull may provoke the release of sham attacks tending to force it to fly up; if this fails the falcon may deftly pick it up from the water surface. A small mammal may release simple straightforward approach and subsequent capturing, &c. Thus we see that the generalized appetitive behaviour was continued until a special stimulus situation interrupted the random searching and released one of the several possible and more specific types of appetitive behaviour. This in its turn was continued until the changing stimulus situation released the swoop, a still more specific type of appetitive behaviour, and this finally led to the chain of consummatory acts.

Baerends (1941) came to the same conclusion in his analysis of the behaviour of the digger wasp *Ammophila campestris* and probably the principle will be found to be generally applicable. It seems, therefore, that the centres of each level of the hierarchical system control a type of appetitive behaviour. This is more generalized in the higher levels and more restricted or more specialized in the lower levels. The transition from higher to lower, more specialized types of appetitive behaviour is brought about by special stimuli which alone are able to direct the impulses to one of the lower centres, or rather to allow them free passage to this lower centre. This stepwise descent of the activation from relatively higher to relatively lower centres eventually results in the stimulation of a centre or a series of centres of the level of the consummatory act, and here the impulse is finally used up.

This hypothesis of the mechanism of instinctive behaviour, though supported by relatively few and very fragmentary facts and still tentative therefore, seems to cover the reality better than any theory thus far advanced. Its concreteness gives it a high heuristic value, and it is to be hoped that continued research in the near future will follow these lines and fill in, change, and adapt the sketchy frame.

NEUROPHYSIOLOGICAL FACTS

The Relatively Higher Levels

The hypothesis presented above, of a hierarchical system of nerve centres each of which has integrative functions of the 'collecting and redispatching' type, has been developed on a foundation of facts of an indirect nature. If it is essentially right, it should be possible to trace these centres by applying neurophysiological methods. As I have said before, it must be considered as one of the greatest advantages of objective behaviour study that by using essentially the same method as other fields of physiology it gives rise to concrete problems that can be tackled by both the ethologist and the physiologist.

Now in recent times several facts have been brought to light which indicate that there is such a system of centres, at least in vertebrates.

I have already mentioned the fact that the work of Weiss, von Holst, Gray, Lissmann, and others proves that the spinal cord of fishes and amphibians must contain mechanisms controlling relatively simple types of co-ordinated movements, such as the locomotory contraction waves of the trunk muscles in fish or the locomotory rhythm of alternating contraction of leg muscles in axolotls. And although doubts have been raised concerning the absolute independence of these centres from external stimulation—doubts which have been discussed in Chapter III—the integrative, co-ordinative nature of the movements controlled by the motor centres is beyond doubt.

Other evidence of the same sort is given by the work of Adrian and Buytendijk (1931) on the respiratory centre in the medulla of fish.

However, all these facts concern the very lowest type of centre we have postulated, that of the consummatory action or, more probable still, that of its least complex component, the fixed pattern.

Now it seems to me to be of the highest importance that Hess (1943, 1944; Hess and Brügger, 1943, 1944; Brügger, 1943) has succeeded, by application of strictly local artificial stimuli, in eliciting behaviour of a much higher level of integration. Hess succeeded in bringing minute electrodes into the diencephalon of intact cats. In this way he could apply weak stimuli to localized parts of the brain. By systematically probing the hypothalamic region he found areas where the application of a stimulus elicited the complete behaviour patterns of either fighting, eating, or sleep. His descriptions make it clear that all the elements of the pattern were not only present but were displayed in perfect co-ordination. Moreover, the response was initiated by genuine appetitive behaviour; the cat looked around and searched for a corner to go to sleep, it searched for food, &c. By combining this experiment with anatomical study the position of the centres of these patterns could be determined (Fig. 90).

These results are of considerable interest in two respects.

First, Hess appears to have found the anatomical basis of the centres controlling instinctive patterns as a whole. A mere electric shock, surely a very simple type of stimulation, releases a complex pattern, an integrated whole of movements of the highest instinctive level. This lends support to our conclusion that somewhere between receptors and effectors there must be a mechanism that takes qualitatively different, configurational impulse-patterns coming from the receptors, combines them in a purely quantitative way, and takes care of redispatching them in reintegrative form so that a configurational movement results. Hess seems to have hit a station somewhere in this mechanism.

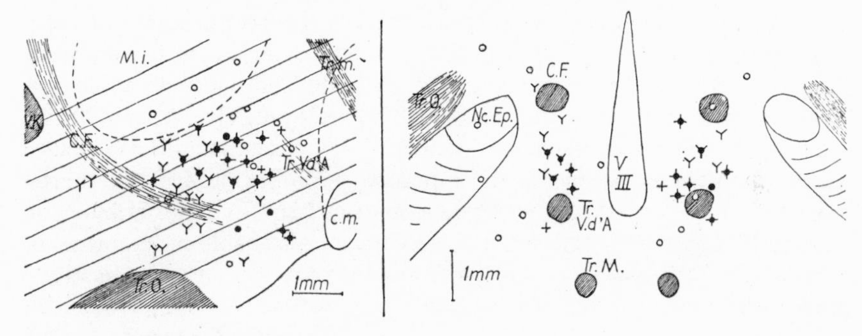

● Food seeking (1)
+ "General motor urge" (2)
Y Fighting in defence (3)
✦ (1) and (2) combined
Y (1) and (3) combined
○ No effect

FIG. 90. Sagittal (left) and horizontal (right) section through hypothalamic 'centres' of a cat. Symbols indicate results of local stimulation. After Brügger, 1943.

Second, the location of these centres is of interest in connexion with the findings about the functions of the spinal cord discussed above. While the spinal cord and the medulla seem to control only certain components of the instinctive patterns, the hypothalamus contains the highest centres concerned with instinctive behaviour. Our analysis of the hierarchical layout of behaviour patterns justifies the prediction that further research along the lines initiated by von Holst, Weiss, Gray, and Hess will lead to the discovery of a whole system of centres belonging to levels below the hypothalamic level as found by Hess, centres which are subordinate to the hypothalamic centres but which in their turn control centres lower still.

I should like to emphasize that this future work could only be done by workers who are fully acquainted with the instinctive behaviour as a whole and with its analysis, and at the same time are in command of neurophysiological methods and techniques. Our science is suffering from a serious lack of students with these qualifications, and it is an urgent task of ethologists and neurophysiologists to join efforts in the training of 'etho-physiologists'.

It is specially interesting that the hierarchical organization has not only been found in vertebrates but in insects as well. According to Baerends's results a wasp with a decentralized system of ventral ganglia and its relatively small 'brain' presents essentially the same picture as vertebrates.

Instinct and instincts. The recognition of the hierarchical organization raises some problems of terminology. There is an enormous confusion around the use of the terms 'instinctive activity' or 'instinctive act'. Some authors maintain that instinctive behaviour is highly variable and adaptive in relation to a goal—in other words that it is purposive or directive—and that, because the goal remains constant while the movements, and hence the mechanisms employed, change, it is futile to attack instinctive behaviour with physiological methods. We have seen that this only applies to the appetitive part of behaviour, and moreover, that even in this purposive element of behaviour the number of possible movements and hence the number of available mechanisms is restricted. Other authors stress the rigidity, the stereotypy of instinctive behaviour.

Now it seems that the degree of variability depends entirely on the level considered. The centres of the higher levels do control purposive behaviour which is adaptive with regard to the mechanisms it employs to attain the end. The lower levels, however, give rise to increasingly simple and more stereotyped movements, until at the level of the consummatory act we have to do with an entirely rigid component, the fixed pattern, and a more or less variable component, the taxis, the variability of which, however, is entirely dependent on changes in the outer world. This seems to settle the controversy; the consummatory act is rigid, the higher patterns are purposive and adaptive. The dispute about whether 'instinctive behaviour' is rigid or adaptive has been founded on the implicit and entirely wrong assumption that there is only one type of instinctive activity.

The fact that the controversy is settled does not, of course, mean that the problem of purposiveness is solved. But the fact that even purposive behaviour appears to be dependent on quantitative activation of a centre and that it comes to an end whenever one of the lower centres has used the impulses shows that purposiveness as such is not a problem which cannot be studied by physiological methods. The fundamental problem is not to be found in the physiological mechanisms now responsible for purposive behaviour but in the history, the genesis of the species.

Returning now to our nomenclatural difficulty, the question naturally arises, What is to be called an instinctive act? Is it the pattern as a whole, or is it one of the partial patterns, or even, as Lorenz has proposed, the consummatory act? I would prefer to apply the name to all levels. For instance, reproductive behaviour in the male stickleback is, as a whole, an instinctive activity. But its component parts, nest-building and fighting, may also be called instinctive activities. A solution could be found by distinguishing instinctive acts of, for example, the first level, the second level, and so on. But here we meet with the additional difficulty that most probably the various major instinctive patterns of a species do not have the same number of levels. If we begin to count from the highest level, we would come to the absurd situation that various consummatory acts, though perhaps of the same degree of complexity, do not belong to the same level. If we begin at the level of the consummatory act, the major instincts would get different rank. This state of affairs renders it impossible to devise a universal nomenclature of instinctive behaviour as long as our knowledge is still in this fragmentary state.

It is of great importance for our understanding of instinctive behaviour as a whole to realize that the various instincts are not independent of each other. We have rejected the reflex hypothesis of behaviour and we have seen that each instinctive mechanism is constantly primed, that is to say, prepared to come into action. Such a system can only work because blocking mechanisms prevent the animal from performing continuous chaotic movements.

Now chaos is further prevented by another principle, viz. that of inhibition between centres of the same level. As a rule, an animal can scarcely do 'two things at a time'. Although there is a certain amount of synchronous activity of two instincts, this is only possible at low motivation, and, as a rule, the strong activation of instinctive behaviour of one kind prevents the functioning of another pattern. Thus an animal in which the sexual drive is strong is much less than normally susceptible to stimuli that normally release flight or eating. On the other hand, when flight is released, the thresholds of the reproductive and feeding activities are raised. The same relationship of mutual inhibition seems to exist between centres of lower levels. Intensive nest-building, for instance, renders the male stickleback much less susceptible than usual to stimuli normally releasing fighting, and vice versa.

Although the physiological basis of this inhibitory relationship will not be discussed here, it should be pointed out that its very existence has been the implicit origin of the distinction between various 'instincts' which has been made by numerous authors. So far, many authors who accepted a distinction between different instincts have defined them in terms of the goal or purpose they serve. A consideration of the neurophysiological relationships underlying instinct leads to a definition of 'an instinct' in which the responsible nervous centres and their mutual inhibition are also taken into account. It makes us realize that the purposiveness of any instinct is safeguarded by the fact that all the activities forming part of a purposive behaviour pattern aimed at the attainment of a certain goal depend on a common neurophysiological mechanism. Thus it is only natural that any definition of 'an instinct' should include not only an indication of the objective aim or purpose it is serving, but also an indication of the neurophysiological mechanisms. Because of the highly tentative character of my picture of these neurophysiological relationships it may seem a little early to attempt a definition of 'an instinct'; yet in my opinion, such an attempt could be of value for future research. I will tentatively define an instinct as a hierarchically organized nervous mechanism which is susceptible to certain priming, releasing and directing impulses of internal as well as of external origin, and which responds to these impulses by co-ordinated movements that contribute to the maintenance of the individual and the species.

For the same reason, it seems too early to attempt an enumeration of the various instincts to be found in animals and man. First, while we know that, in the cat, eating, fighting, and sleep must each be called a major instinct because each is dependent on the activation of a hypothalamic centre, there are patterns which almost certainly are equally dependent on a relatively high centre (e.g. escape, sexual behaviour, &c.) but of which nothing of the kind has yet been proved. Further, different species have different instincts. For instance, while many species have a parental instinct, others never take care of their offspring and hence probably do not have the corresponding neurophysiological mechanisms. However, such things are difficult to decide at present, because, for instance, it has been found that males of species in which the care of the young is exclusively an affair of the female can be brought to display the full maternal behaviour pattern by injecting them with prolactin (see p. 66). Though this example concerns individuals of the same species, we could not reject *a priori* the possibility that, for instance, a species might lack a certain instinct because, having lost it relatively recently, it retained the nervous mechanism but not the required motivational mechanism. So long as we know nothing about such things, it would be as well to refrain from generalizations.

However, it is possible to point out some inconsistencies in the present views on instincts to be found in the literature. Contrary to current views, there is, in my opinion, no 'social instinct' in our sense. There are no special activities to be called 'social' that are not part of some instinct. There is no such thing as the activation of a system of centres controlling social activities. An animal is called social when it strives to be in the neighbourhood of fellow members of its species when performing some, or all, of its instinctive activities. In other words, when these instincts are active, the fellow member of the species is part of the adequate stimulus situation which the animal tries to find through its appetitive behaviour. In some species all instincts, even the reproductive instinct and the instinct of sleep, have social aspects. In many other species the social aspect, while present in feeding or in all non-reproductive instincts, is absent from the reproductive instinct. This is especially obvious in many fishes and birds. In many amphibians the situation is just the reverse. Further, in many species there are differences of degree, or even of quality, between the social elements of different instincts. For instance, in herring gulls there is a tendency to nest in colonies. But in mating and nest-building there is only a weak social tendency, limited to the fact that individuals select their nesting site in the neighbourhood of an existing colony; attacking a predator, one of the other sub-instincts of the reproductive instinct, is a much more social affair..

There is no instinct for the selection of the environment, no *Funktionskreis des Milieus* as von Uexküll (1921) claims. Here again reactions to habitat are parts of the reproductive instinct or of other instincts.

There is, however, an instinct of sleep. Sleep is a readily recognizable, though simple behaviour pattern and has a corresponding appetitive behaviour pattern; further, it is dependent on the activation of a centre. Moreover, sleep can appear as a displacement activity (see below), a property found in true instinctive patterns only.

There is, further, an instinct of comfort, or rather of care of the surface of the body.

There is not one instinct of combat. There are several sub-instincts of fighting. The most common type of fighting is sexual fighting, which is part of the reproductive pattern. Sexual fighting has to be distinguished from defence against a predator, for it has a different IRM and, often, a different motor pattern.

Displacement activities. Another set of interrelations, though in itself perhaps not of primary importance in the organization of behaviour, has to be considered now: those revealed by the occurrence of 'displacement activity'. This phenomenon will be discussed in some detail, because it is not generally known and yet seems to be of great importance for our understanding of the neurophysiological background of instinct.

It has struck many observers that animals may, under certain circumstances, perform movements which do not belong to the motor pattern of the instinct that is activated at the moment of observation. For instance, fighting domestic cocks may suddenly pick at the ground, as if they were feeding. Fighting European starlings may vigorously preen their feathers. Courting birds of paradise wipe their bills now and then. Herring gulls, while engaged in deadly combat, may all at once pluck nesting material, &c. (Fig. 91). In all the observed instances the animal gives the impression of being very strongly motivated ('nervous'). Rand (1943) has called such movements 'irrelevant' movements. Makkink (1936) gave an implied interpretation by using the term 'sparking-over movements', suggesting that impulses are 'sparking over' on another 'track'. Kirkman (1937) used the term 'substitute activities' which I adopted in 1939 (Tinbergen, 1939). Later the term 'displacement activity' was proposed (Tinbergen and Van Iersel, 1947; Armstrong, 1948), and this term will be used here.

The phenomenon has been clearly recognized and analysed independently by Kortlandt (1940*a*) and by Tinbergen (1939, 1940). An examination of the conditions under which displacement activities usually occur led to the conclusion that, in all known cases, there is a surplus of motivation, the discharge of which through the normal paths is in some way prevented. The most usual situations are: (1) conflict of two strongly activated antagonistic drives; (2) strong motivation of a drive, usually the sexual drive, together with lack of external stimuli required for the release of the consummatory acts belonging to that drive.

1. A conflict between two antagonistic drives is found with animals fighting at the boundary line between their territories. Numerous instances, apart from those already mentioned, have been observed. Male sticklebacks, when meeting at the boundary between their territories, adopt the attitude seen in the picture serving as a frontispiece for this book. At first sight the movements seem very similar to feeding movements, and so they were at first described (Tinbergen, 1940). It has since been discovered, however, that with very strong motivation, when the movement becomes more complete, it develops into complete digging, and cannot be distinguished from the movements of digging a pit for the nest. This movement does not belong to the fighting drive but to the nesting pattern.

It is a striking fact that displacement activities often occur in a situation in which the fighting drive and the drive to escape are both activated. Within its own territory, a male invariably attacks every other male. Outside its territory, the same male does not fight, but flees before a stranger. In between the two situations, that is, at the territory's boundary, opposing males perform displacement activities. The natural conclusion, viz. that displacement activities, in this situation, are an outlet of the conflicting drives of attack and escape (which of course cannot discharge themselves simultaneously, because their motor patterns are antagonistic), has been tested experimentally (Tinbergen, 1940). A red dummy was offered to a male stickleback in its territory and was duly attacked. Instead of withdrawing the dummy, it was made

FIG. 91. Various displacement activities.

1. Nesting movements in herring gull as an outlet of the fighting instinct. Moderate intensity. After Tinbergen, 1940.
2. Nesting movements in herring gulls as outlets of the fighting instinct. High intensity.
3. Sleeping attitude in European oystercatcher as an outlet of the fighting instinct.
4. Sleeping attitude in European avocet as an outlet of the fighting instinct. After Makkink, 1936.
5. Sand-digging in male three-spined stickleback as an outlet of the fighting instinct. After Tinbergen, 1947*b*.
6. Preening in sheldrake as an outlet of the sexual instinct. After Makkink, 1931.
7. Preening in the garganey as an outlet of the sexual instinct. After Lorenz, 1941.
8. Preening in the mandarin as an outlet of the sexual instinct. After Lorenz, 1941.
9. Preening in the mallard as an outlet of the sexual instinct. After Lorenz, 1941.
10. Preening in the European avocet as an outlet of the sexual instinct. After Makkink, 1936.
11. Food-catching movement in the European blue heron as an outlet of the sexual instinct. After Verwey, 1930.
12, 13, and 14. Sexual movements in the European cormorant as outlets of the fighting instinct. After Kortlandt, 1934*b*.
15. Food-begging movements in herring gull as outlets of the sexual instinct. After Tinbergen, 1940.
16. Food-pecking movements in domestic cocks as an outlet of the fighting instinct.

to 'resist' the attack by hitting the attacking male with it. When this 'counter-attack' is carried out vigorously enough, the territory-holding male can be defeated in its own territory. It withdraws and hides in the vegetation. If the dummy is now held motionless in the territory, it will continuously stimulate the male's fighting drive. The tendency to flee, however, diminishes with time. Gradually the fighting drive regains its superiority over the tendency to flee, and after a few minutes the male will attack the dummy again. Just before this happens, however, the male performs displacement digging. This shows, therefore, that displacement digging occurs when the two drives involved are in exact equilibrium. There is little doubt that the various displacement activities occurring during territorial fights must be explained in the same way.

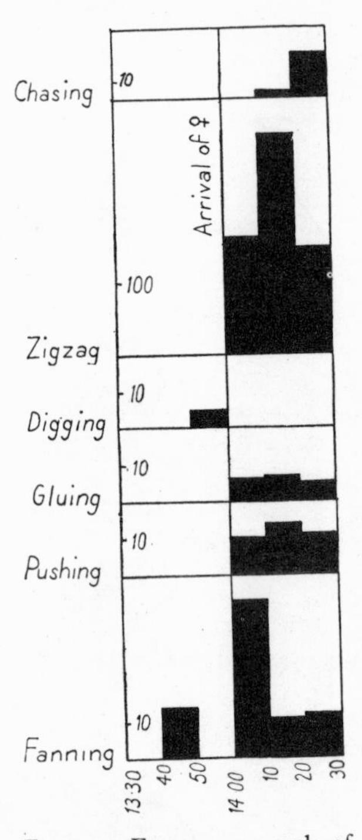

FIG. 92. Frequency graph of reproductive activities in male three-spined stickleback before and after visual stimulation by a female. Abscissa: time in hours and minutes. After Tinbergen and Van Iersel, 1947.

2. In many species the male, even when strongly sexually motivated, is unable to perform coition as long as the female does not provide the sign stimuli necessary for the release of the male's consummatory act. A male stickleback, for instance, cannot ejaculate sperm before the female has deposited her eggs in the nest. The appearance of a female strongly arouses his sexual impulses. When the female does not respond to his zigzag dance by following him to the nest—a common phenomenon in incompletely motivated females—the male invariably shows nest-ventilating movements, often of high intensity and of long duration (Fig. 92). The amount of displacement fanning can even be used as a very reliable measure of the strength of his sexual motivation (Van Iersel and Tinbergen, 1947). This doubtless is the reason why so many displacement activities are parts of courtship patterns. Male ducks, for instance, regularly preen their plumage during courtship; birds of paradise, as mentioned above, and also European jays, wipe their bills. Herring gulls and many other birds (see Lack, 1940) feed their mates during courtship; it is not at all improbable that this courtship feeding is displacement feeding. Numerous other examples have been mentioned in the literature (Kortlandt, 1940; Tinbergen, 1940; Lorenz, 1941).

A comparative review of displacement activities reveals that they are always innate patterns, known to us from the study of other instincts. Recognition is often difficult because the displacement activity is usually incomplete. This turns upon the intensity of the motivation, for with very strong motivation the displacement act may be complete. An

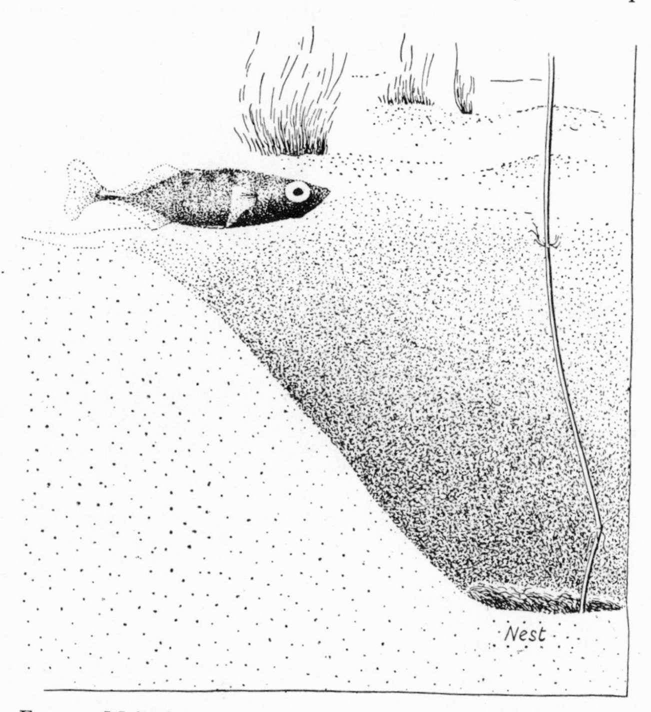

FIG. 93. Male three-spined stickleback with nest in deep pit made by displacement digging. After Tinbergen and Van Iersel, 1947.

example is given by Tinbergen and Van Iersel: if male sticklebacks are forced to nest very closely together, they will show nearly continuous displacement digging and the result is that their territories are littered with pits, or even become one huge pit (Figs. 93, 94). Another cause of difficulty in recognizing displacement activities is the 'ritualization' which a certain group of them secondarily undergo, a phenomenon which will be discussed later (p. 192). The remarkable stereotypy of displacement reactions, the fact that they resemble innate motor patterns of other instincts, and the fact that they are typical for the species and do not differ from one member to another, suggests that the motivation of an instinct when prevented from discharging through its own motor pattern finds an outlet by discharge through the centre of another instinct. The facts known thus far are well in accordance with the 'centre-theory' of instinct presented above and elaborated below (p. 122). The incompleteness of displacement activities shows that the sideways discharge meets with considerable resistance.

It is a very remarkable fact that our present knowledge of displacement activities confirms our conclusion that care of the body surface, and also sleep, are true instincts depending upon special nervous motivational (excitatory) centres, for they both appear as displacement

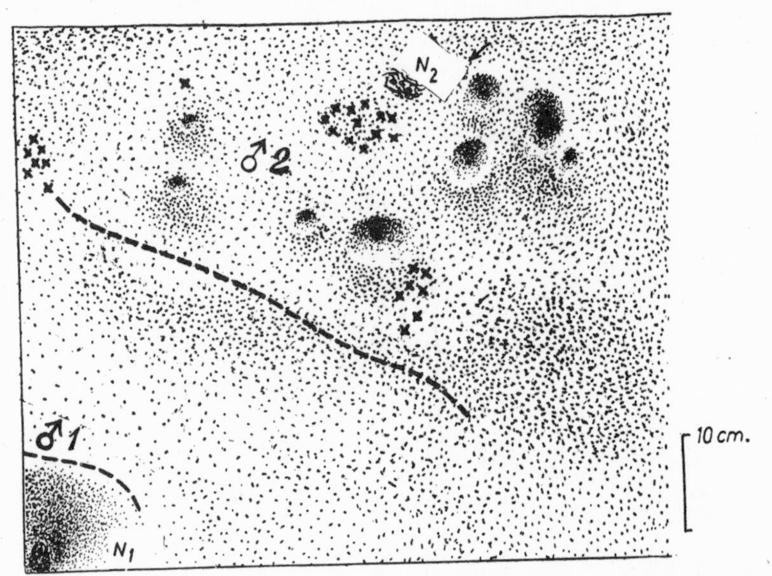

FIG. 94. Territories of two neighbouring male three-spined sticklebacks in crowded tank, with pits due to displacement digging. Broken lines: boundaries where owners are threatening; the ground between them is no-man's-land. Crosses indicate *Elodea* and *Batrachium* vegetation. N_1, nest belonging to male 1 (the male of Fig. 93); N_2, nest belonging to male 2. After Tinbergen and Van Iersel, 1947.

reactions. It may be recalled that sleep, like instinctive consummatory acts, has a special kind of appetitive behaviour directed towards attaining a fitting situation (p. 113); further that sleep, like true instinctive activities, depends on the activation of a centre in the hypothalamus (p. 113). In these respects, therefore, sleep behaves as an instinctive activity, and there is no doubt, in my opinion, that sleep should be called an instinct. On the other hand, there are, so far as we know, no instances of displacement activities that could be interpreted as 'social activities' or as 'activities in relation to the environment', and this confirms our conclusion that there is no such thing as 'a social instinct' or a *Funktionskreis des Mediums*.

As will be clear later, displacement activities are by no means rare in man. Many instances of 'nervousness' concern displacement activities. The situation in man, however, is more complicated than in animals.

The fact that a displacement activity is an expression, not of its 'own' drive (of autochthonous motivation, as Kortlandt (1940) called it) but of a 'strange' drive (of allochthonous motivation), makes it possible for it to act as a signal to fellow members of the same species, provided it can be distinguished from the 'genuine' activity, activated by its 'own' drive. As a matter of fact, many displacement activities are different from their 'models' and do act as signals. Thus displacement digging in sticklebacks is actually understood by other sticklebacks as a threat (as an expression of the fighting drive); it is different from its 'model', true digging, in that the spines are erected. The fact that displacement reactions may in this way embark upon an evolutionary development of their own by acquiring a social function will be discussed further below.

The Lower Levels

The principle of hierarchical organization has been studied from a different point of view in the lower levels of integration. As we have seen, the consummatory act is the lowest or simplest element that appears in most cases of overt behaviour. It is, however, obvious that in most consummatory acts we have to do with co-ordinated movements of a great number of muscles. Now many facts show that there is hierarchical organization within each consummatory act too.

The ventilation or fanning movement of the stickleback may serve as an example. It consists of alternating forward and backward movements of the pectoral fins and, synchronous with their rhythm, sinusoid swimming movements of trunk and tail. The unity of this movement as a whole breaks down under certain conditions, e.g. when the fish is diseased. The pectorals may still move in alternation, but the trunk moves in an independent rhythm.

The movements of each separate fin are, as a rule, a co-ordinated system of swinging movements of a number of separate fin rays. Each ray swings to the left and the right in a regular alternating rhythm, and between successive rays there is a definite phase difference. Von Holst (1934) found in the goldfish that under certain conditions of anaesthesia the co-ordination within each fin may be disturbed; the rays begin to flutter in disorder. However, he reports that even then the movements of each separate ray remain perfectly regular, each ray swinging in perfect left-right alternation; it is merely the co-ordination between the rays that is disturbed. In these cases, therefore, it appears that the consummatory act is a co-ordinated activity of a number of fins, each of which may, under certain conditions, become independent. The movements of the fin as a whole are composed of still lower units which may also act independently of each other. One step farther down brings us on the level of the separate muscle contractions, for each fin ray is moved by a set of two muscles to the right and two to the left, the left and right pair contracting in alternation, and each pair acting as one muscle.

The argument for a hierarchical organization in behaviour as a whole, from the level of a major instinct down to the consummatory act, is different from that for the lowest levels. As we saw above, the existence of mutually inhibiting relations between different reactions (between the centres of the same level)—which enables the animal to confine its

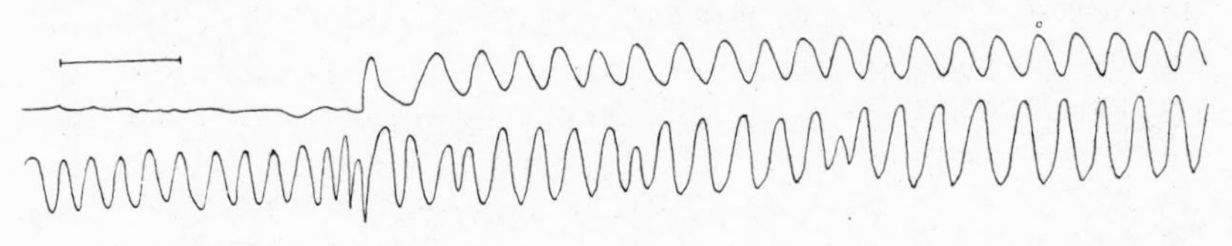

FIG. 95. The 'magnet effect'. Registration of the movements of the pectoral fin (dominant, above) and the dorsal fin (dependent, below) of *Labrus*. *a–b*, the dominant rhythm is at rest; the dependent rhythm appears in pure form. *b–c*, the dependent rhythm follows the dominant rhythm's tempo incompletely. *c–d*, the dependent rhythm is completely 'caught'; increase in amplitude is caused by superposition. After von Holst, 1936.

activities to one thing at a time—makes it possible to distinguish between more or less independent units within each level. Within the consummatory act, on the other hand, there is co-operation instead of inhibition, and the independence of component parts becomes apparent in quite another way, viz. when they occasionally follow their own rhythms under conditions that break down the power of the co-ordinating agents. While it is beyond my power to give an exhaustive review of the work done on integrative mechanisms at these lower levels, I want to mention the important work done by von Holst. This author made a thorough analysis of the co-ordinating principles at work in the swimming movements of fishes. As we have seen, the separate fins usually move in strict alternation or synchronization. Von Holst proved, with the aid of a wealth of material, that this co-ordination was due to two principles. First, the rhythm of one centre, for instance that of the pectoral fins, may be superimposed upon the rhythm of another centre such as that of the caudal fin. This superposition effect may vary over a wide range of intensities, from nearly zero to practically absolute dominance (Fig. 62, p. 73).

The second type does not concern the intensity of the separate contractions but their frequency or rhythm. In many cases pectorals and caudal fins have different rhythms. Usually one of them is dominant and, in some unknown way, can force the other to fall into step (Fig. 95).

This influence too may be of varying intensity, and von Holst shows many cases ranging from weak to strong dominance. In cases of slight dominance the co-ordination is not absolute but relative, and von Holst shows that absolute co-ordination is but an extreme case of relative co-ordination.

The nature of this influence is unknown.

A few remarks may be inserted here on the configurational character of many processes involved in behaviour. Configurational processes may be found both on the receptor and on the motor side of behaviour. At the receptor side, the evidence suggests that in the eye the retina itself has synthesizing powers. Moreover, simple motor responses display configurational qualities. Although the configurational character of perceptual processes has often been stressed in a descriptive way, no attempt has been made, so far as I know, to analyse 'configurational' stimuli and the inter-relationships responsible for them. This is because correlation of ethological and neurophysiological findings is, at present, more promising at the effector side than at the receptor side of behaviour. The reason for this is that neurophysiology is much more developed in the domain of motor response than in that of receptor processes. 'The laboratory usage for obtaining reflexes is often direct stimulation of bared afferent nerves, a plan which eschews selective excitation of specific receptors and precise knowledge of the receptive field, and thus renounces serviceable guides to the functional purposes of the reflex' (Creed, Denny Brown, Eccles, Liddell, and Sherrington, 1932, p. 104).

In the domain of the motor response, von Holst (1941) has shown that the phenomenon called the magnet effect, which in itself seems to be open to quantitative description, may unite two rhythms into a movement of a higher order which displays all the characteristics of a configurational process, in which relations (between the tempi of dominant and dependent rhythm) are more constant than quantities (the absolute tempi). This is, so far as I know, the first attempt to attack the problem of *Gestalt* along physiological lines.

P. Weiss's Concept of Nervous Hierarchy

The concept of a hierarchical organization of the nervous system is, of course, not new. And it is especially interesting to see how ethological study has led to the recognition of the hierarchical structure of innate behaviour quite independently of the conclusions drawn by neurophysiologists.

Now the ethologist has been considering higher levels of integration than the neurophysiologist. As a result, a combined picture of neurophysiological and ethological facts shows more levels than those recognized by neurophysiologists.

Weiss (1941*a*) enumerates the following levels from the lowest upward.

1. The level of the individual motor unit.
2. All the motor units belonging to one muscle.
3. Co-ordinated functions of muscular complexes relating to a single joint.
4. Co-ordinated movements of a limb as a whole.
5. Co-ordinated movements of a number of locomotor organs resulting in locomotion.
6. 'The highest level common to all animals', the movements of 'the animal as a whole'. (Weiss, 1941*a*, p. 23).

The levels 3, 4, and 5 are those studied by von Holst in his work on co-ordination in fishes. As will be clear, level 6 in Weiss's scheme really consists of a number of levels, in fact all the levels from the 'fixed pattern' up It is interesting that Weiss's classification stops just here, because it is just at this level that one type of co-ordination changes into another type. But it will be clear that the hierarchical principle is the basis of the organization of these higher levels too.

Here again is an illustration of the fundamental identity of the neurophysiological and the ethological approach. The only difference between them is a difference in level of integration.

CONCLUSION

To conclude this section on the physiology of instinctive behaviour, it would be of advantage to present the results obtained in the form of a graphic picture of the nervous mechanisms involved. I should like to emphasize the tentative nature of such an attempt. While such a graphic representation may help to organize our thoughts, it has grave dangers in that it tends to make us forget its provisional and hypothetical nature.

We have seen that the causal factors controlling innate behaviour are of two kinds, viz. internal and external. In most cases both kinds exert an influence and they supplement each other. Usually the internal factors do not themselves evoke the overt response; they merely determine the threshold of the response to the sensory stimuli. Therefore, the internal factors like hormones, internal stimuli, and intrinsic impulses determine what the psychologist calls the motivation; and I will call them motivational factors. As we have seen, it is highly probable that in many cases external stimuli may also raise the motivation, and some of them therefore also belong among the motivational factors.

Another category of external stimuli, viz. those activating releasing mechanisms, must be distinguished from the motivational factors; I shall call them releasing factors.

Beach (1942), in a discussion of the factors effective in arousing the male sexual behaviour in rats, postulated a central excitatory mechanism (CEM) which is receptive to sensory stimuli and hormone influences

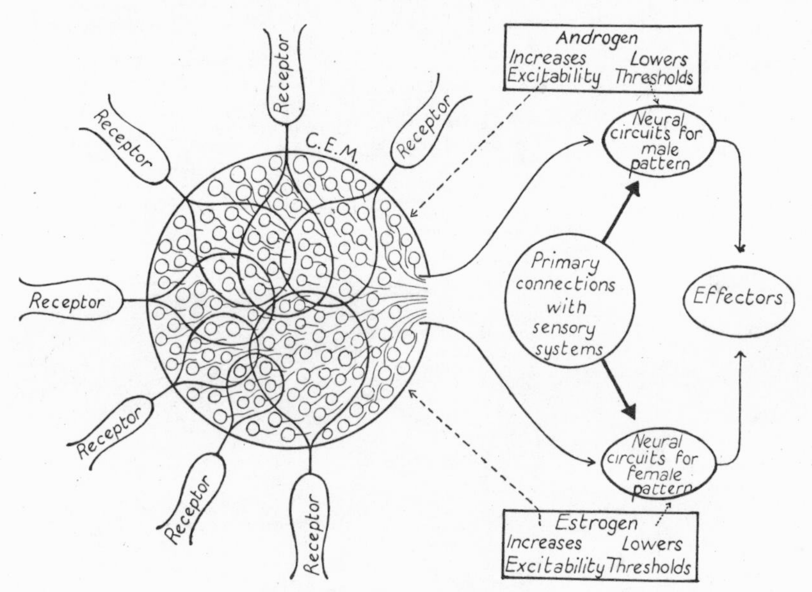

FIG. 96. The 'central excitatory mechanism', after Beach, 1942.

and which dispatches impulses to the neural circuits of the behaviour pattern.

When we compare his presentation (Fig. 96) with the results discussed above, it is clear that, although Beach gives due attention to the co-operation of internal and external factors, his picture does not take into account (1) the hierarchical organization and (2) the different functions of motivational and releasing factors.

Concerning the first point, it seems that the single CEM postulated by Beach is rather a system of CEMs of different levels. Each 'centre' in our system is a CEM in Beach's sense, as each of these centres has its own afferent and efferent connexions.

Of no less importance is the difference between motivational and releasing factors. For, as we have seen, the motivational factors influence the CEM itself while the releasing factors activate a reflex-like mechanism, the IRM, removing a block that prevented the outflow of impulses along the efferent paths.

The system C(entral) E(xcitatory) M(echanism)—I(nnate) R(eleasing)

M(echanism) is tentatively presented in Figs. 97 and 98. Let us first consider Fig. 97, which represents one centre of an intermediate level.

The centre is 'loaded' by motivational impulses of vaiious kinds. First it receives impulses from the superordinated centre of the next higher level. Impulses from this higher level flow to other centres as

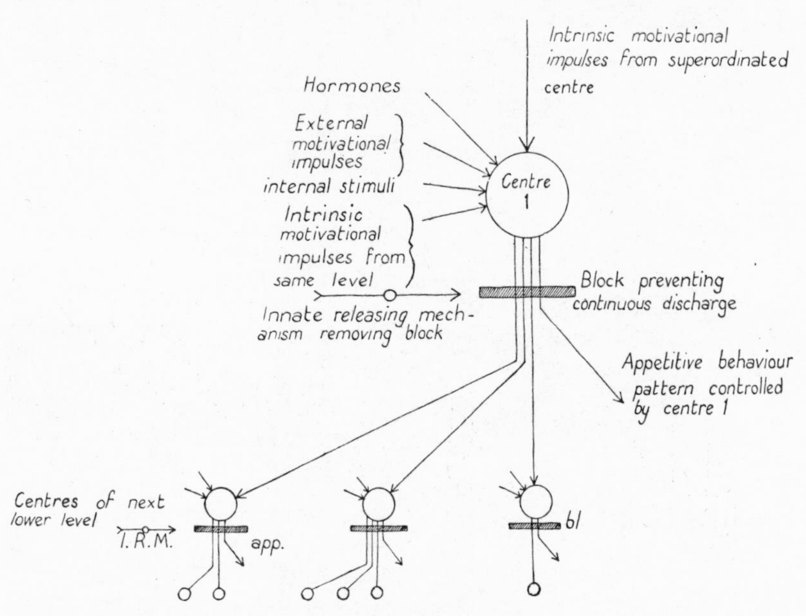

FIG. 97. Tentative representation of an instinctive 'centre' of an intermediate level. Explanation in text.

well, in fact to all the centres controlled by the higher centre. Second, centre 1 may receive impulses from an 'automatic', self-generating centre belonging especially to it (see p. 73 where the dual nature of centres as found by von Holst is discussed). Third, a hormone might contribute to the motivation, either by acting directly on centre 1, or through the automatic centre. As discussed on p. 74, it is probable that hormones act exclusively on the higher centres. Fourth, internal sensory stimuli (see p. 66) may help to load centre 1. Fifth, external sensory stimuli might also act directly upon the centre and contribute to its motivation.

This system together represents a CEM in Beach's sense, belonging to one level of the hierarchical system.

Outgoing impulses are blocked as long as the IRM is not stimulated. When the adequate sign stimuli impinge upon the reflex-like IRM, the block is removed. The impulses can now flow along a number of paths. All but one lead to subordinate centres of the next lower level. However, all these centres are prevented from action by their own blocks, and most of the impulses therefore flow to the nervous structures controlling appetitive behaviour. This appetitive behaviour, as we have seen (p. 106), is carried on until one of the IRMs of the lower level removes a

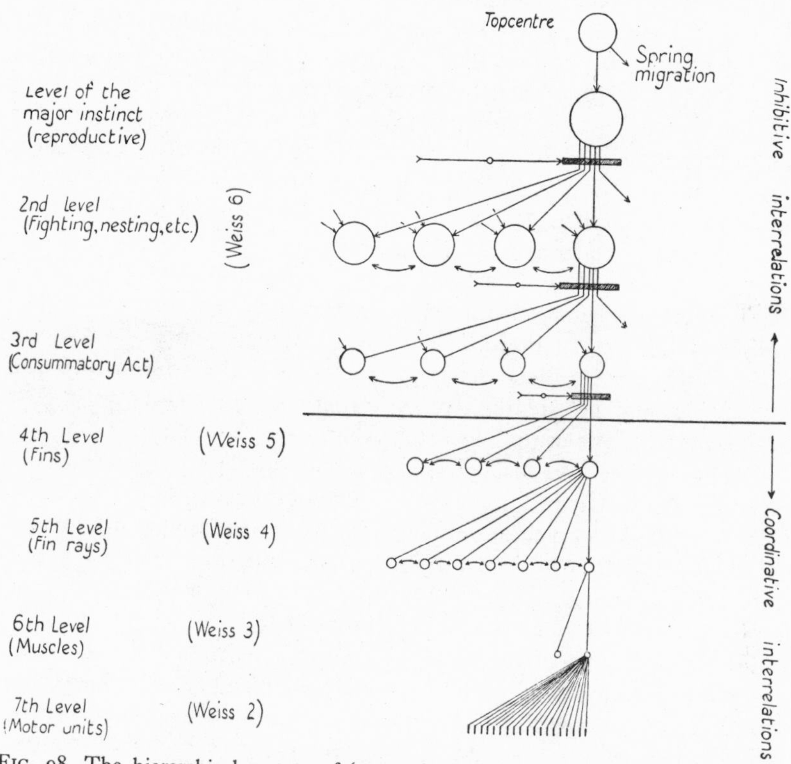

FIG. 98. The hierarchical system of 'centres' underlying a major instinct, viz. the reproductive instinct of the male three-spined stickleback. Explanation in text.

block, as a result of which free passage is given through the corresponding centre of this next lower level. This 'drains away' the impulses from the appetitive behaviour mechanism and conducts them to the appetitive behaviour mechanism of the lower centre.

Fig. 98 suggests how centres of this type might be organized within one major instinct. The reproductive instinct of the male three-spined stickleback has been taken as a concrete example. The hormonal influence, presumably exerted by testosterone, is acting upon the highest centre. This centre is most probably also influenced by a rise in temperature. These two influences together cause the fish to migrate from the sea (or from deep fresh water) into more shallow fresh water. This highest centre, which might be called the migration centre, seems to have no block. A certain degree of motivation results in migratory behaviour, without release by any special set of sign stimuli, which is true appetitive behaviour. This appetitive behaviour is carried on—the fish migrates—until the sign stimuli, provided by a suitable territory (shallow warm water and suitable vegetation) act upon the IRM blocking the reproductive centre *sensu stricto*, which might be called 'territorial centre'. The impulses then flow through this centre. Here, again, the paths to the subordinated centres (fighting, nest-building, &c.) are blocked as long as the sign stimuli adequate to these lower levels are not forthcoming. The only open path is that to the appetitive behaviour, which consists of swimming around, waiting for either another male to be fought or a female to be courted, or nest material to be used in building.

If, for instance, fighting is released by the trespassing of a male into the territory, the male swims towards the opponent (appetitive behaviour). The opponent must give new, more specific sign stimuli, which will remove a block belonging to one of the consummatory acts (biting, chasing, threatening, &c.) in order to direct the impulse flow to the centre of one of these consummatory acts.

The various centres located at the various levels are, therefore, not organized in exactly the same way. The very highest centre has no block. If there were blocks at these very highest centres, the animal would have no means of 'getting rid' of impulses at all, which, as far as we know, would lead to neurosis. The next centre responds, in comparison to the lower centres, to a relatively higher number of motivational factors.

The next centres of the male stickleback's reproductive behaviour pattern, represented, for example, by fighting and nest-building, are loaded primarily by the impulses coming from the higher centre. Whether there are special motivational factors for each of these centres besides those coming from the higher centre is not certain, but I think there are, because the fighting drive displays the phenomenon of after-discharge, and, moreover, seems to be motivated (not merely released) by external stimuli. In general it seems, that the lower we go, the more pronounced the influence of external releasing stimuli becomes.

Arrows between the centres of one level indicate interrelationships suggested by the existence of mutual inhibition and of displacement activities. It should be emphasized that it is quite possible that these interconnexions do not in reality run directly from one centre to the other, but go by way of the superordinated centre, and that the 'inhibition' of one centre by the other may in reality be competition, the 'inhibiting' centre by decrease of resistance 'draining away' the impulse flow at the expense of the 'inhibited' centre.

Thus the motivation is carried on through a number of steps, which may be different in different instincts, down to the level of the consummatory action. Here the picture is changed. When the block of the consummatory centre is removed, a number of centres come into action simultaneously, between which horizontal co-ordinative forces are effective. With fish, these centres below the consummatory level are arranged in two or three planes, the lowest of which is the centre of the right or left fin-ray muscle, which is a relatively simple type of nervous centre. The relation between sub-consummatory centres of the same level are represented by horizontal lines.

It should again be emphasized that these diagrams represent no more than a working hypothesis of a type that helps to put our thoughts in order.

VI

THE DEVELOPMENT OF BEHAVIOUR IN THE INDIVIDUAL

As I said in the introductory chapter, the study of development cannot be separated from the study of the functions of the adult animal; it is part of our central problem of the causation of behaviour. Although there are records of an overwhelming number of detailed facts bearing on the problem of behaviour development, a consistent synthetic treatment of the evidence is not yet possible. Yet it will be of advantage to consider some of the evidence in the light of the views developed in the foregoing chapters.

As far as innate behaviour is concerned, development can be considered complete when the animal reaches maturity. But it should be realized that this stage is not reached by all behaviour patterns at the same time. Most of them, usually the non-reproductive patterns of the adult individual, complete their development at an early stage, whereas the reproductive patterns appear much later. Moreover, they subsequently regress and redevelop repeatedly, at least in many species, and this phenomenon deserves no less attention than the development of the other activities.

Further, although this book is concerned with innate behaviour, learning processes cannot be ignored entirely. The main reason for this is that all learning effects a change in the innate functions; learning, therefore, is a phenomenon of ontogenetic growth of behaviour superimposed on the innate patterns and their mechanisms. Now although it is a continuation of the growth of innate patterns, many learning processes occur so early in the developmental history of the individual that they often precede the completion of innate patterns.

Another and more important reason for including learning in this chapter is the fact that there is a close relationship between innate equipment and learning processes, in that learning is often predetermined by the innate constitution. Many animals inherit predispositions to learn special things, and these dispositions to learn therefore belong to the innate equipment.

THE GROWTH OF INNATE MOTOR PATTERNS

The study of development of behaviour is still in its infancy. Nevertheless what scanty information we have is sufficient to see the main trend of the problems immediately ahead.

First of all, it is our task to find out to what extent the changes in the behaviour during individual development are merely due to growth and to what extent they are due to learning processes. Critical studies on this problem have been carried out at the locomotion level by a small number of authors. The work of Carmichael (1926, 1927) on the swimming movements of tadpoles, that of Weiss (1941*a*) on walking in Urodeles, and that of Grohmann (1939) on flight in birds will be taken as examples. Carmichael's work is based on the well-known fact that tadpoles carry out incipient swimming movements while still in the egg capsule. These movements improve in the course of days, partly during the life in the egg, partly during early life as a frog larva. In order to test whether this improvement in function was the effect of learning or of growth, Carmichael reared a group of eggs under continuous chloretone anaesthesia. When a control group of the same age, reared simultaneously under normal conditions, reached a stage at which locomotion could be judged to be fully developed, the chloretone-treated larvae were placed in fresh water and thus in the same environment as the controls. They showed locomotion of the same degree of accomplishment as the controls. This shows that tadpole locomotion is not learnt: it is an innate mechanism revealed in the course of growth.

In a different way, Weiss came to the same conclusion about the ambulation of Urodeles. In numerous transplantation experiments, in which antagonistic muscles of a leg were detached from their own tendons and were caused to unite with each other's tendons, while their nerve-supply was left intact, he found that each muscle reacted according to its original function. This resulted, of course, in limb movements just the opposite to those occurring in normal animals. On a higher level of co-ordination, that of the fore limb as a whole, Weiss got the same result by interchanging right and left fore limb rudiments; since their antero-posterior axes are predetermined, the limbs so grafted now faced backwards instead of forwards. When the nerve connexions had been established, the grafted limbs moved just as they would have done had they been left in their original position, causing backward motion when the rest of the animal was trying to move forward, and forward motion when the rest of the animal was trying, for example, to avoid a noxious stimulus presented in front of it (Figs. 99, 100). A year's experience did not change this reversed movement of the grafted legs.

Still another method led to the same conclusion. When the whole spinal cord of the tadpole was deafferentated at an early stage, preventing the growing larva from getting sensory information about its movements, the basic patterns of co-ordination still developed. The contradictory conclusions of Detwiler and Vandyke (1934) and of Chase (1940) seem, according to Weiss, to be based on secondary effects due

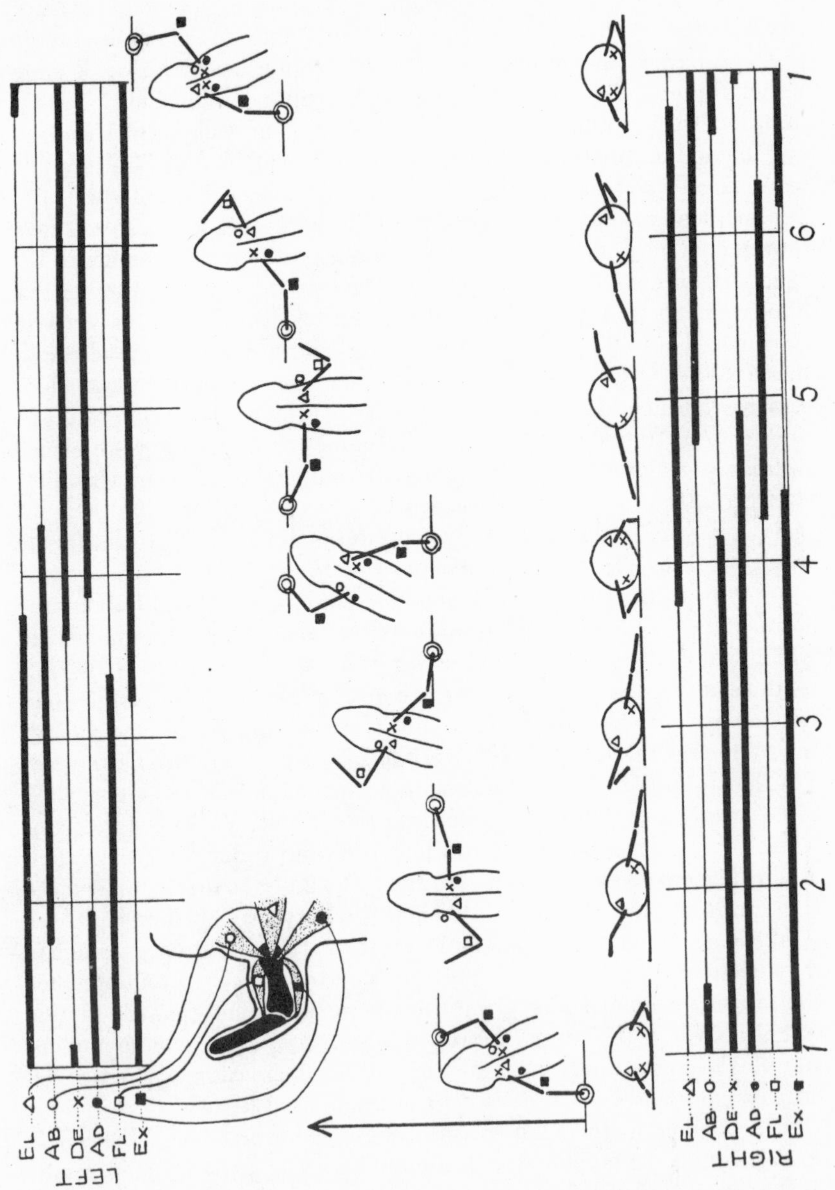

Fig. 99. Time record of the fore-limb action of a salamander during ambulatory progression over solid ground. Numbers 1–6 represent successive stages. Heavy lines indicate periods of action of the various muscles. *El*, elevator; *Ab*, abductor; *De*, depressor; *Ad*, adductor; *Fl*, flexor; *Ex*, extensor. After Weiss, 1941*a*.

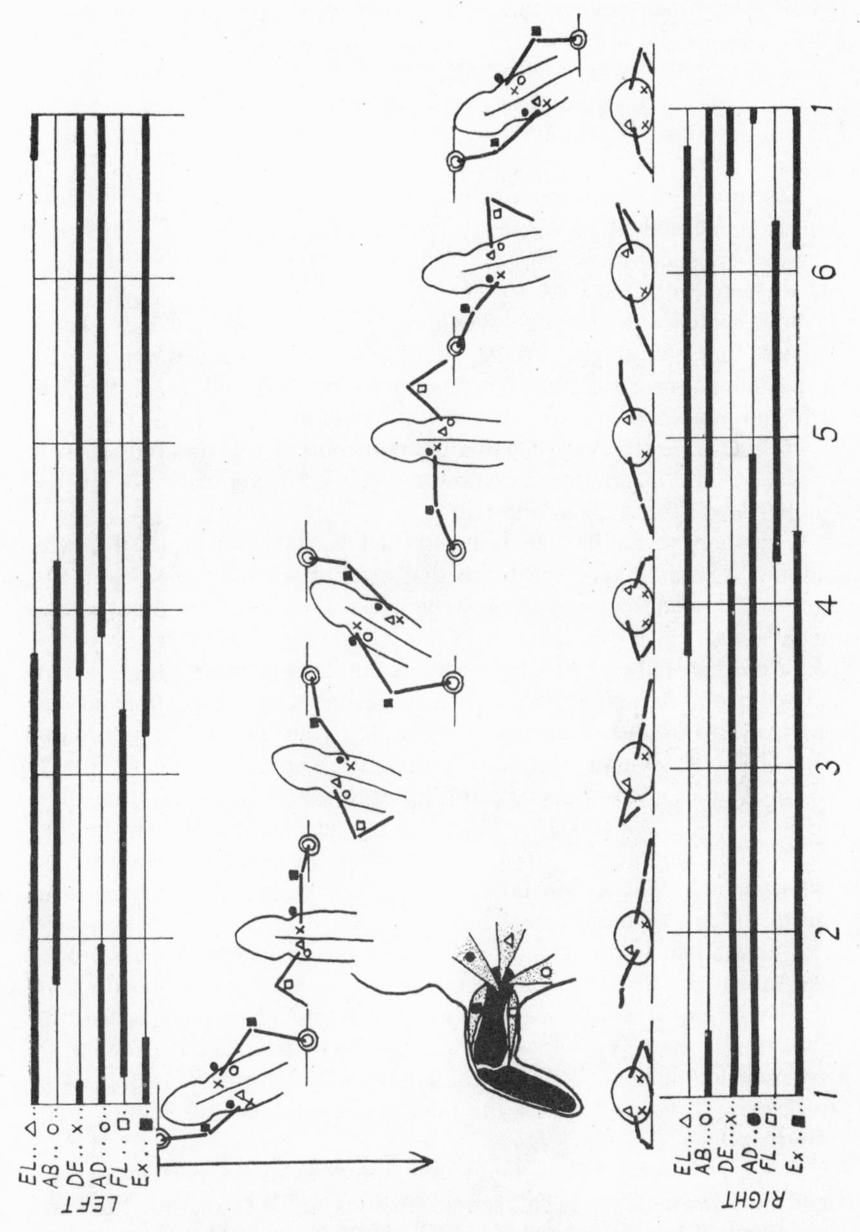

FIG. 100. Time record of the fore-limb action of a salamander with interchanged, i.e. reversed, fore limbs during ambulation. For explanation see Fig. 99. After Weiss, 1941*a*.

to the experimental conditions, slight damage to the spinal cord resulting in a chronic hypertonicity of the whole motor pool involved. In anuran tadpoles, where extirpation of spinal ganglia and dorsal roots can be carried out without causing additional damage to the spinal cord, Weiss got much clearer evidence of the self-differentiation of the locomotory patterns (Weiss, 1941*a*).

Essentially the same picture was found with the aid of the tendon-crossing method in rats (Sperry, 1941). But here the first signs of a regulatory function of experience were visible, especially in the behaviour of the fore limbs (Sperry, 1942).

The evidence in the higher mammals is scarce, but what we know shows that, although certain adaptations and readjustments after experimental or accidental transpositions of this sort do occur, they are effected in a very slow and clumsy way, even in man.

It should be stressed that this phenomenon of self-differentiation is typical of the locomotory and lower levels of movement only. In the higher levels the story is different.

The flight of birds has been studied by Grohmann (1939), who applied a method very similar to that used by Carmichael. Grohmann prevented young pigeons from carrying out their incipient flight movements not by anaesthesizing them but by rearing them in narrow tubes, thus mechanically preventing them from moving their wings. When the controls were able to fly a certain distance, both experimental and control birds were submitted to a simple flying test and it was found that the achievements of the two groups were equal.

I do not know of systematic studies of this kind in invertebrates, but in both butterflies and dragonflies I have repeatedly watched the first flight after hatching and found that when the animals are disturbed before they have made any incipient flight movements they may perform perfect flights of hundreds of metres.

The gradual improvement of locomotory movements appears to be dependent not primarily on growth in either receptors or muscles, but largely on growth of the nervous system. It should be emphasized first that this tentative conclusion is a guess based on relatively few facts; second, that our knowledge of the enormously complicated complex of processes called 'growth of the nervous system' is still highly fragmentary.

A parallel study of developmental processes in the nervous system and in behaviour has been carried out by Coghill (1929) for the early stages of behaviour in the axolotl larva. Coghill observed that the earliest appearance of certain movements occurred simultaneously with the appearance of certain nervous connexions which, by their topography, were thought to mediate these movements. Thus, the first movement is a reflex flexure of the anterior trunk muscles opposite to the stimulated side. With increasing age, the area of muscle involved extends farther tailward, until at last the whole trunk responds by 'coiling' (Fig. 101). Now at the time when the first flexure movement was performed, neurological examination detected the first cross-connexion between the

FIG. 101. The 'coil' of an *Amblystoma* larva. After Coghill, 1929.

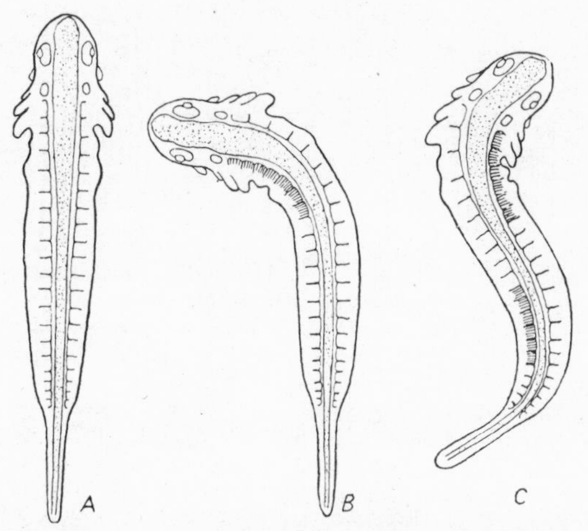

FIG. 102. A. Diagram of *Amblystoma* of the early swimming stage. B. The first flexure in *Amblystoma* of the early swimming stage. Contracting muscle-segments cross-hatched. C. The swimming movement of *Amblystoma* of the early swimming stage. The first flexure has passed tailward and the second flexure is beginning in the anterior region. After Herrick and Coghill, 1915, from Coghill, 1929.

sensory and the motor paths, the so-called floor-cells (Fig. 103). The gradual tailward extension of the flexure coincides with a spreading of the development of the motor cells from the front, where development begins, to the rear.

At the next stage the body can be bent into an S. At this stage a coil begins in the anterior region and spreads tailward, but before the contraction has reached the tail a new, reversed flexure appears at the head (Fig. 102).

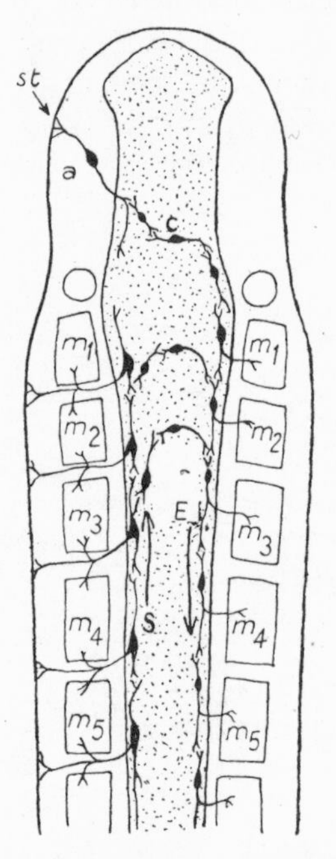

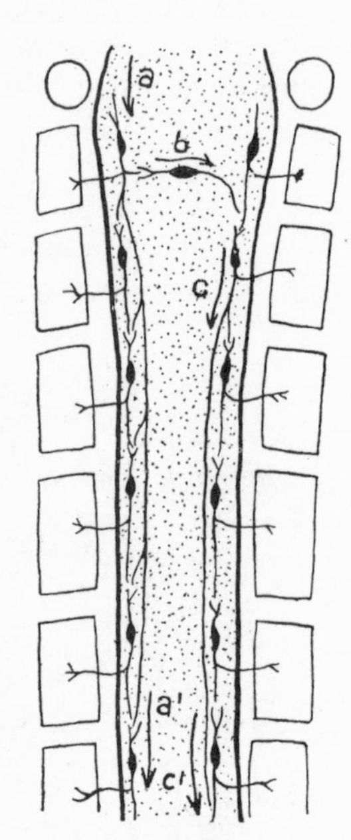

FIG. 103. Left: A diagram of the mechanism which accounts for the coil reaction, cephalocaudal progression of movement, and movement away from the side stimulated in *Amblystoma*. A stimulus in front of the ear (*St*) would excite an afferent neurone (*a*), which, in turn, would excite a commissural cell (*c*) in the floor plate. This cell turns caudad in the motor tract of the other side. A stimulus behind the ear, whether in the skin or in the muscle, excites the afferent neurones which also excite the commissural cells of the floor plate, through which the impulse reaches the motor tract (*E*) of the other side. Arrow *S* indicates sensory path, arrow *E* indicates motor path. m_1, $_2$, &c., indicate muscle-segments. Brain and spinal cord stippled.

Right: A diagram of the neuromotor mechanism of swimming in *Amblystoma*. The sensory mechanism is omitted. The arrows indicate the direction of conduction. Arrow *a* represents the initial impulse which, as it passes tailward to *a′* and beyond, excites the muscle-segments to a wave of contraction that progresses tailward. By the time the animal can swim, these neurones of the motor tract in the anterior region have developed collaterals which grow towards the median plane into synapse with commissural cells of the floor plate. This relation is illustrated at *b*, where the arrow indicates an impulse passing to the motor system of the other side. In the motor system it passes tailward according to the arrows *c* and *c′*. After Coghill, 1929.

The appearance of this S flexure coincides with the development of connexions between the sensory cells and the motor cells of the same side.

Although Coghill's interpretation of the swimming rhythm is based on the chain reflex scheme, which is now considered an incomplete and, in a way, inadequate picture (see Chapters III and V), his observations must of course be accepted, and they show that nerve growth and development of behaviour go hand in hand.

It is not necessary to elaborate this point further; we may safely assume that development of overt behaviour usually is the result of growth processes in the nervous system and not of growth of muscles or receptor organs, which, as a general rule, are developed earlier. So far as is known, only the very earliest movements of developing embryos are usually myogenic.

However, there are cases of quite another category where development of overt behaviour is not dependent on development of either nervous system or receptor and effector organs, but on the increase of hormone stimulation. This phenomenon will be discussed below.

Before proceeding to describe the development of behaviour in the different levels of integration, the problem of the development of the innervation pattern should be briefly mentioned, because the latest results obtained in this field throw an interesting light on the problem of the localization of centres discussed in Chapter III.

Weiss (1941*a*) extended his tendon-crossing experiments (p. 129) by combining them with nerve crossings. He succeeded in growing connexions between flexor nerve and extensor muscle in a limb, and vice versa. This was done both after tendon crossing and without tendon crossing. The general result was that the nerve, after regeneration, transmitted the impulse functionally appropriate to the muscle it now innervated. That is to say, when the central nervous system sent out a 'flexor call', the original extensor nerve carried the impulses to the flexor muscle, while the original flexor nerve did not transmit impulses. In this way Weiss showed that the responsiveness of a motor nerve is not dependent on its localized connexion with the centres but is, in some yet unknown way, conditioned to selective responsiveness by the muscle it innervates. In his first publications Weiss stated this conclusion, which in my opinion cannot be doubted, in combination with a hypothesis about the nature of this 'conditioning' influence of the muscle. The hypothetical nature of his discussion of this second problem has caused scepticism about the first conclusion as well. As Weiss points out quite clearly in his 1941 paper, the results of the nerve-crossing experiments should not be discarded because of uncertainty about the way in which the muscle brings about the specific moulding of its motor innervation.

These facts show that our picture of the localization of centres must not be too precise and formalistic. There is a gross anatomical localization of, for instance, locomotory centres in the spinal cord, and of higher centres in the hypothalamic region, but the finer localization is of a physiological and topographically variable nature.

Returning now to the description of the course of ontogenetic development, we again have to rely on Coghill's classic observations on *Amblystoma*. As I have already mentioned, the movements in the first stages of development are simultaneous, mass movements of the trunk somites. The general course of further development is that these mass movements differentiate into patterns, the components of which gradually become distinct from the behaviour as a whole. A differentiated pattern of locomotion does not arise by construction, by addition of parts, but in just the reverse way, by differentiation of a diffuse whole. For instance, the limbs in *Amblystoma* at first are only capable of moving synchronously with, and in the same way as, the neighbouring trunk somites. Later they are able to move relatively independently. In a comparable way, gill movements are at first completely dependent on trunk movements; they become capable of independent movements later on.

The differentiation of feeding behaviour follows another course. The complete feeding behaviour of the larva consists of a forward leap, accompanied by snapping and subsequent swallowing. The order of appearance is: first the leap is shown alone, later actual snapping is added, while swallowing is added later still. This type of development could properly be called addition rather than differentiation.

Although there are many scattered reports about the appearance of certain reactions as a whole—such as feeding, escape, &c.—analyses of such movements together with a study of the incipient stages such as that presented by Coghill are seldom given. From what we know of the order of appearance of reactions, it appears that each species has a fixed time pattern just as with growing morphological structures. A fine instance of a study of such a time pattern is Mrs. Nice's work on the development of behaviour of the song sparrow (Nice, 1943).

As an instance of analytical work in this field, the thorough study made by Kortlandt (1940) of the European cormorant may be cited. Kortlandt studied the ontogenetic development of several behaviour patterns, and found, contrary to what we could expect from a generalization of Coghill's results on the locomotion pattern, that the component parts mature independently and are later combined into purposive patterns of a higher order. Thus complete nesting behaviour consists of fetching twigs, pushing them into the nest, and fastening them there by the typical quivering movement shown by so many birds. The young begin to show nesting behaviour at an age of 2 weeks. This consists merely of quivering, which is performed quite 'senselessly', for the twigs are not fastened. When 4 or 5 weeks old, the young continue quivering until the twig gets caught by the nest. Still later they accept twigs from the male and also go to fetch twigs for themselves to work them into the nest.

Our analysis of instinctive behaviour allows us to describe this in more general terms: the lower units, of the level of the consummatory acts, appear first and the appetitive behaviour appears later.

A similar process seems to occur in the development of the pattern of feeding the young. The complete pattern consists of regurgitating and then opening the mouth, bending over the young and allowing it to put its head into the mouth. Young of 2½–3 weeks of age show the first incipient feeding behaviour without the introductory regurgitating. Here again the consummatory act appears first and the introductory behaviour follows much later.

It would seem to me that no general descriptive rules of the ontogenetic development of instinctive behaviour can be established until we know a good deal more. It would be of especial importance to give attention not only to the appearance of individual components but also to the development of pattern in relation to that of its component parts. It seems that one cannot generalize Coghill's conclusion that individual acts of behaviour 'crystallize out' from a diffuse total response, and that a kind of additive type of integration may play a part too, perhaps especially in the higher levels. In this respect it is certainly of considerable interest that the development of *Amblystoma's* feeding reaction, being a response of a higher level than the locomotion responses, seems to follow a course more similar to that described by Kortlandt than to that found by Coghill for locomotion.

THE MATURATION OF INNATE MOTOR PATTERNS

The problem of the growth of neural motor patterns should not be confused with the problem of seasonal maturation of the reproductive patterns. For, although in both cases non-experimental observation reveals a gradual development of overt behaviour, we know with relative certainty that in the arousal of reproductive patterns no growth of the nervous mechanisms is involved; they are ready for action the whole year round. One indication is the fact that the reproductive patterns can be activated out of season by administration of sex hormones, and, as Lashley has pointed out (1938), these reactions seem to be too quick to be accounted for by a trophic influence of the sex hormones on nerve-cells. There is, however, a still stronger indication. Sexual reactions may in some species appear as displacement reactions providing an outlet for impulses from other instincts, such as fighting. Now Kortlandt found in the European cormorant that, although sexual behaviour is not shown during autumn and early winter or at best is at a low ebb, displacement reactions derived from the sexual pattern but motivated by the fighting drive are shown the whole year round, and neither their frequency nor their intensity fluctuates with the sexual instinct. This important discovery shows plainly that the nervous mechanism is present the whole year round and that the fluctuations in autochthonous sexual behaviour are a matter of hormone fluctuations. For this reason it is necessary to distinguish between growth and maturation.

The development of behaviour patterns under the influence of hormones has some interesting aspects. First, there is the unmistakable fact that the hormone acts on the higher instinctive centres. In the stickleback, fighting and nest-building are activated simultaneously, and although experimental work on this problem is still in a provisional stage, there is little doubt that the male sex hormone is responsible for the activation of both, that is to say, of the superordinated centre.

Within these patterns, the course of the maturation process is marked by the sequence in which the partial patterns appear. In the stickleback, there is a temporal sequence of the various components within the nest-building subinstinct. For instance, digging a hole is activated before actual building. This is probably due to the fact that the digging centre has a lower threshold for the impulses from the nest-building centre than the building *sensu stricto* and the other components; so that, when the motivation is still low, all impulses will 'flow' to the digging centre.

It is an interesting fact that digging is also the only displacement reaction through which the fighting drive can find an outlet. If our conception of the causation of a displacement reaction is right (that is to say, if a displacement reaction is caused by a collateral escape of impulses from one centre to another centre of the same level), it is strange that obstruction of the fighting drive does not result in the activation of the complete nesting pattern, but of only one component of it. But if digging is the component of this pattern that is most easily activated because it has the lowest threshold for impulses coming from the nesting centre as a whole, it is understandable that it should be the only component 'used' as displacement reaction for the dissipation of the impulses coming from the blocked or thwarted fighting centre.

This one instance will make it clear that an understanding of the sequence in which the partial patterns of an instinct appear can only be gained after a thorough analysis of the pattern involved.

The maturation of later stages of the reproductive instinct may involve changes other than merely in the level of one hormone. For instance, the nest-building phase in the stickleback is followed by a mating stage, the only stage during which the male is ready to fertilize the eggs. It is not known whether this stage is initiated by a change in hormone level or by a qualitative hormonal change, or perhaps by an external stimulus supplied by the finished nest.

After having fertilized a number of clutches, the male loses its sexual drive and begins to take care of the eggs. This again may be dependent on external stimuli and/or on a hormonal change. During the egg stage and the subsequent stage of caring for the young there are again several changes in behaviour. Here, too, external changes may influence behaviour, as they are actually known to do in one instance: the increase in frequency of fanning or ventilation is dependent on the increasing oxygen consumption of the eggs, as has been proved experimentally. It is, however, also dependent on internal factors; one is inclined to think of a hormone change.

These few remarks may suffice to show how complex the phenomena of maturation are; so little is known about them, however, that it is impossible to interpret them in terms of general rules.

Descending to a still lower level, such as that of the digging pattern, we find that their maturation follows still another course. The digging response matures gradually. In its complete form it involves swimming towards the nesting site, pointing the head downward, thrusting the snout into the bottom, sucking up sand, swimming away with a mouthful and spitting it out 10–20 cm. away.

At the outset, when motivation is lowest, digging consists of a scarcely discernible downward bending of the head. The next stage consists of the actual boring of the snout into the soil without sucking up sand. In the next stage the animal sucks sand and drops it on the spot. Then, with increasing intensity, the last component, that of carrying the sand away, is gradually added, the distance increasing from a few centimetres up to 20 and sometimes more (Fig. 104). Thus the development of digging offers a quite clear picture: the first link of the chain of movements appears first in time and then the whole chain is gradually completed (1) by addition of subsequent elements, and (2) by intensification of each element, the earlier elements achieving their final form before the later elements. This again could be explained by assuming a different threshold for each of the elements of the chain.

Nest-building in birds shows a similar development. According to the observations of Howard (1920), the first indication of the awakening of the nesting drive in female buntings is an occasional picking up of a piece of nesting material, which is, however, dropped instantaneously. With growing motivation, the bird will carry it about for some seconds before dropping it. Later still, it carries the material to a nesting site but drops it there, sometimes after having made some half-hearted building movements. In this way, the action chain is completed gradually.

Another instance is the development of copulation as can be seen in the herring gull. Here again it is the first elements in the chain that

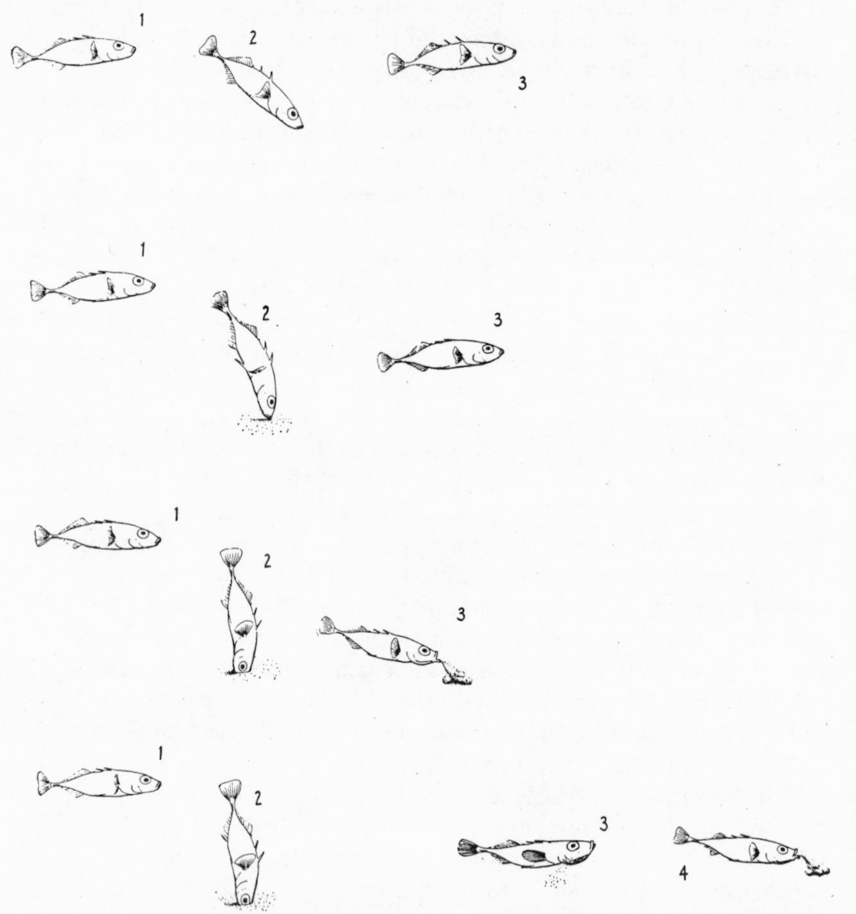

FIG. 104. Sand-digging in the male three-spined stickleback in four intensities. Top: lowest intensity. Bottom: highest intensity. Numbers 1, 2, 3, 4 indicate successive phases of one movement. Further explanation in the text.

mature first and the subsequent elements are added in regular order until the behaviour is complete.

Finally, the work of Kortlandt on the development of behaviour in the cormorant gives us an opportunity to compare the maturation of behaviour in the adult bird with the growth in the young. Kortlandt emphasizes the fact that the growth of behaviour patterns in the young usually begins with the specialized stereotyped components at or near the end of a chain and that the introductory elements, comprising the appetitive behaviour, are added retrogressively, i.e. in the reverse order to that of their final performance, as in the nesting behaviour. Maturation in the adult is, according to Kortlandt, 'just the reverse', the generalized appetitive behaviour being performed first, and then the reaction is completed by extension towards the more specialized, subordinated appetitive behaviour until the pattern is completed by the appearance of the consummatory act.

So far as I know, maturation of this type is a phenomenon of general occurrence. With the increase of motivation, the elements of a behaviour sequence develop in the order in which they are eventually performed; appetitive behaviour appears first, and the acts which complete the behaviour appear last. Each element appears first in the low intensity, and the order of the increase of intensity follows the order of appearance.

FIG. 105. Intention movement of male wheatear (*Oenanthe oenanthe*).

This rule, here illustrated by examples of increasing motivation due to an increasing hormone concentration, seems to hold good for other motivational factors too. Whenever motivation crosses its threshold value, the first reactions to maximal adequate external stimulation are the initial parts of the chain, and these appear in incomplete form. The experienced observer can often make a rather accurate guess about what instinct is being motivated because he recognizes those incipient movements. Heinroth (1911), who was the first to call attention to them, accordingly called them *Intentionsbewegungen*, which could be rendered by 'intention movements'.

In many cases an intention movement is an incomplete locomotory movement. In birds, for instance, a growing motivation to fly is often indicated by incipient flying movements, such as flattening the plumage, stretching the neck, bending the tarsal joint (Fig. 105), &c.

As locomotion is often part of the appetitive behaviour of several different instincts, an incipient locomotory movement can only indicate that the animal is about to go somewhere, but which one of the various instincts is being motivated cannot be judged. However, the direction is often indicated too. Thus a labyrinthine fish in which the air-snapping instinct begins to work—which happens every few minutes—shows its intention by lifting its body so that it points obliquely upward, moving no more than perhaps 2 cm. in the direction of the surface. After a few seconds it may move a little again, then wait again, and after a varying number of these initiatory movements it will suddenly dash to the surface and take in air (Fig. 106).

It is clear that this picture of the growth and the maturation of motor response is very fragmentary. Although the distinction between growth and maturation made in this chapter seems to be based on firm ground, our further analysis is very tentative. More work is urgently needed.

LEARNING PROCESSES

As this book deals with innate behaviour, it is not its task to provide anything like an exhaustive treatment of learning. However, the views

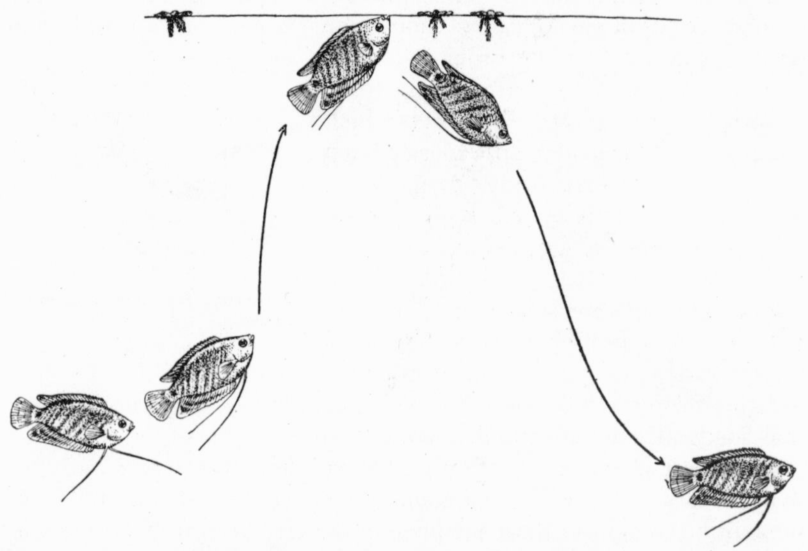

FIG. 106. Air-snapping and its intention movement in the labyrinth fish, *Colisa labiosa*.

on the physiology of instinct developed in the foregoing chapters may throw light on some aspects of the problem of learning. The student of innate behaviour has, moreover, to know something of learning processes in order to distinguish them from growth or maturation processes. Also, it might be useful to approach learning phenomena from a more naturalistic standpoint than is usually done and to ask, not what can an animal learn, but what does it actually learn under natural conditions?

It is extremely difficult to give an exact objective definition of learning. This is due to several circumstances. First, the term comprises a great variety of phenomena. Second, since man has to learn much during individual life, and since most of his learning processes are known to him by introspection, man tends to approach the problems of learning more from the subjective side than he does when studying unconditioned behaviour. For our purpose it will do to use the following provisional definition: learning is a central nervous process causing more or less lasting changes in the innate behavioural mechanisms under the influence of the outer world.

The Detection of Learning

First, some words should be said on the detection of learning processes, since both for the student of innate behaviour and for the student of conditioning they are often difficult to tell apart from other phenomena.

Every kind of learning brings about a change in behaviour during individual life. This change may concern the form of a movement, i.e. the execution of the motor response itself, or, more often, it is the releasing mechanism that is changed, while the motor response remains unaltered.

Growth and maturation. In tracing evidence of learning one meets with several difficulties. Where motor responses themselves are changed during individual life, one is readily inclined to ascribe the change to learning. This conclusion—a generalization of what is at least partially true of man—is, however, too bold.

As we have seen, growth and maturation processes may give rise to changes that give the observer the impression of a learning process. We have seen that the gradual improvement of the swimming movements in tadpoles and of flight in young birds is primarily due to growth and not to learning. However, this seems so utterly improbable to the naïve observer that, if experimental proof were lacking, most people would not hesitate to consider them clear examples of learning.

It is still more difficult to imagine the possibility of changes due to growth in cases where, by contrast to the examples mentioned, it is not the motor response that is changing but its dependence on external stimulation. In a study of the response of young ducks and geese to models representing birds of prey (p. 32) it was found that ducklings showed escape reactions to the optimal models at a very young age, whereas goslings did not react to these models, but only to the alarm call of the parents. When, at an age of several weeks, the goslings began to react to the dummies themselves, the natural conclusion would be that they had been conditioned to them. However, rearing the goslings in isolation revealed that the sensitivity to the dummy appeared at the same age; this proved that the sensitivity was innate but that it developed much later than in ducks.

Sympathetic induction and imitation. A number of misinterpretations are brought about by the phenomenon of 'sympathetic induction'. In social species every individual tends to behave like its companions. Katz and Révész (1909) showed that fowl kept in isolation, after being fed until they did not accept food any more, began to eat again as soon as they saw other birds eating. This phenomenon has nothing to do with learning through imitation; the motor responses of pecking food are completely innate in fowl, or at least are not learned by imitation. The uniformity of behaviour is brought about by induction, one bird activating the feeding instinct in the other to such a degree that the external stimuli, which were below threshold value, are now sufficient to release the pecking response.

The ability to learn motor responses by imitation is remarkably restricted. It is highly developed in man and the higher mammals; in other groups I know of only one instance, viz. the song of some songbirds. According to Heinroth's observations, male skylarks, nightingales, chaffinches, goldfinches, and several other species have to learn their song by imitation.

Acquisition of skill. The conclusion drawn from the experiments of Carmichael, Grohmann, and Weiss on growth of the motor patterns of locomotion, viz. that the improvement observable in the performance during individual development is due to growth, must not be taken too rigidly. The flight of young birds, for instance, shows a slow and often subtle improvement during the months following the period over which Grohmann's experiments were carried out: the movements of alighting, of manœuvring in a strong and changeable wind, of stalking or hunting prey, &c., while innate in a general way, are undoubtedly changed by a kind of learning process called acquisition of skill. It is not known in any concrete case what part of the development of such a motor pattern is due to growth and what part to learning, and the only way to find out is to extend experiments over longer periods. For instance, it is well known that preventing young birds from flying for a much longer period than that covered by Grohmann's experiments retards them appreciably in the development of skill in comparison with normal individuals (see also Thorpe, 1951).

Raising in isolation. The only way to know what animals do learn under natural conditions is to raise them, to study the development of their behaviour, and to test by experiment whether observed changes are due to growth or to learning. In general, such tests may be of two kinds: one way is to exclude a possible influence of the environment; the other way is to attempt to change the behaviour experimentally by deliberate changing of environmental elements and see how that affects behaviour.

Individual differences as indications of learning. A pretty reliable indication of learning is to be found in differences between the behaviour of individuals. Because learning is dependent on individual experiences in relation to the outer world, and because the external situations supplying the material for learning are so variable, the chances are that no two individuals learn exactly the same things. For instance, when one individual chick responds to the farmer's voice by running towards him while other chicks do not, it is probable that the first chick has been conditioned to it.

In wild animals, however, much learning occurs in circumstances that are more or less identical for every individual of a species or at least of a population, and because of this fact the converse of the above is not true: identity of behaviour of many individuals by no means indicates that their behaviour is entirely innate. For instance, many European songbirds have to learn that wasps of the genus *Vespa* are inedible. Inexperienced individuals capture a wasp as readily as a fly. The repellent taste of a wasp's entrails and, in some cases, the wasp's sting, usually condition them on the very first occasion and teach them to avoid wasps, and, in fact, all flies of similar colour as well (Mostler, 1935). As these experiences have an effect lasting at least several months, every adult individual of the species avoids wasps and their 'mimics'. As was mentioned already, Heinroth has shown that many male songbirds have to learn their song from males of the same species. Young nightingales, for instance, do not develop their typical song when reared in isolation and they imitate any members of other species which have been singing in their neighbourhood. Wild males, however, having been raised by their own species, all have the song of the species, and without experiments it would be impossible to know that the song has to be learned.

The Innate Disposition to Learn

'Localized learning.' The student of innate behaviour, accustomed to studying a number of different species and the entire behaviour pattern, is repeatedly confronted with the fact that an animal may learn some things much more readily than others. That is to say, some parts of the pattern, some reactions, may be changed by learning while others seem to be so rigidly fixed that no learning is possible. In other words, there seem to be more or less strictly localized 'dispositions to learn'. Different species are predisposed to learn different parts of the pattern. So far as we know, these differences between species have adaptive significance.

Some instances may illustrate this important fact of localized dispositions.

Herring gulls have a number of innate reactions to the young: they brood them, feed them, and rescue them if attacked by strangers or predators. Interchanging the young of two nests of the same age has very different effects, depending on the age of the young. When they are only a few days old they will be accepted by their 'foster parents'. But if the same test is made when the young are more than 5 days old, they will not be accepted. This means that after a period of about 5 days, during which a parent herring gull is willing to take care of any young of the right age, the parents are conditioned to their own young. They will then neglect or even kill any other young forced upon them (Tinbergen, 1936). Approximately the same results have been obtained in various species of terns (Watson and Lashley, 1915). This learning to 'know' the chicks individually is very remarkable, for the human observer rarely succeeds in distinguishing the young and never reaches the same degree of accuracy as the birds.

FIG. 107. Herring gull in choice test, hesitating between its own and another gull's eggs.

The ability of a herring gull to learn its own eggs is, by contrast, amazingly poor. The eggs of different gulls vary a good deal in colour and speckling, in fact they vary much more than the chicks do. Yet even gulls that have eggs of a very distinctive type such as bluish, poorly pigmented eggs, or eggs with exceptionally large or small spots, never show any preference for their own eggs (Fig. 107). The innate releasing mechanism of the brooding reactions does not undergo any change by conditioning, so far as the egg itself is concerned. There is, in this respect, a sharp contrast between the reactions to young and those to eggs (Tinbergen, 1936).

The sexual pattern, again, is readily conditioned. Herring gulls, like a great many other birds, are strictly monogamous, and each bird confines its sexual activities to its own mate once the formation of pairs has taken place. Here again the gull's ability to recognize its mate is far superior to our powers of recognizing the gulls. There is proof of the amazing fact that a herring gull instantly recognizes its mate (that is, reacts selectively to it amongst a group of other gulls) from a distance of 30 yards. Nor is the herring gull alone in this respect; similar facts are known about jackdaws, geese, terns, and other birds. Recognition is based partially on visual stimuli, partly on voice.

The fact that many species, man included, seem to distinguish individuals of their own species much more readily than individuals of other species is another aspect of the innate basis of learning.

Other instances of localized learning dispositions have been found in the digger wasp, *Philanthus triangulum*. Females of this species have innate releasing mechanisms directing the chain of prey-hunting acti-

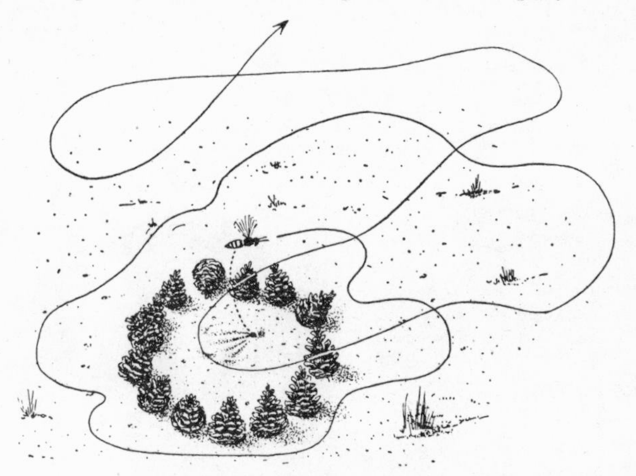

FIG. 108. Locality study of a digger wasp, *Philanthus triangulum*.

vities to the hive bee alone, among hundreds of other insect species. There is no indication of a conditioning of the hunting pattern, apart, perhaps, from the development of a certain preference for favourable hunting territories. Each wasp, however, learns, with astonishing rapidity and precision, the locality of each new nest it builds. It has been proved experimentally that the so-called locality studies (Fig. 108), which have also been observed in numerous other Hymenoptera, are the means of learning the position of the nest in relation to certain landmarks. The best achievement has been observed in the following case (Tinbergen and Kruyt, 1938). A ring of twenty pine-cones was put around a nesting-hole while the wasp was inside. Upon leaving, she made a locality study lasting 6 seconds, and then left. The pine-cones were thereupon taken away and deposited a foot away in the same arrangement. When the wasp returned after about 90 minutes with a captured bee, her choice between the real nest and the displaced beacons was watched thirteen times, the beacon being displaced each time (Fig. 109). She chose the beacons in all the tests and only found the nest after the original situation had been restored. Thus, in a great number of varied experiments, it has been found that the homing ability of *Philanthus* and of other digger wasps depends on an amazing learning capacity (Tinbergen, 1932; Tinbergen and Kruyt, 1938; Baerends, 1941; Van Beusekom, 1948). In strong contrast to this, neither the hunting behaviour nor the food-seeking pattern shows any trace of conditioning. This is especially striking when we compare *Philanthus* with the hive bee. Both species suck nectar, but while the honey bee gets readily conditioned to special species of plants and special localities, nothing of the kind could be observed in *Philanthus*. While the homing capacities

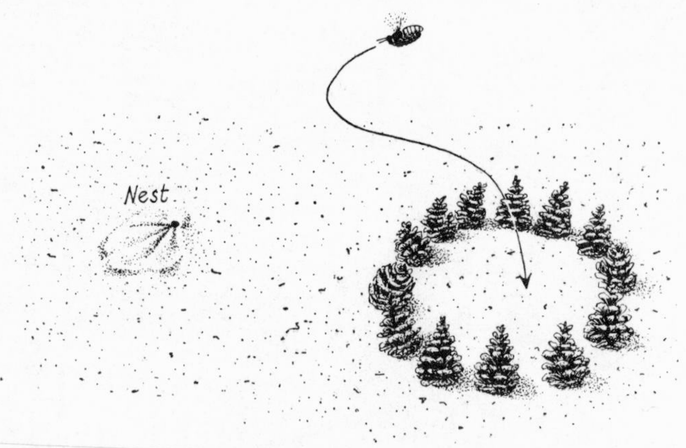

FIG. 109. Orientation experiment. Explanation in the text.

in the two species are very similar and involve about equal learning capacities, they make a very different use of their learning abilities in the food-seeking instinct.

Another type of this complex interweaving of innate and learned capacities has been reported in jackdaws. In a jackdaw community, as in so many other bird colonies, many relationships between members are based on individually acquired attachments and antagonisms. One such relationship based on learning is the so-called social order. Each individual learns to avoid some of its colony-mates which are stronger or more aggressive, and learns to know which of them can be treated in an unfriendly manner without risk. Now Lorenz (1931) found that when a female that occupies a relatively low position in the social order gets mated with a high-ranking male, her position is at once rated just as high as that of her mate by all the members of the colony. These amazing learning processes are found in the very jackdaws that show such poor learning ability in relation to either eggs or young.

Ravens starting to build nests for the first time in their lives are highly unselective in choosing sticks to build them with. However, they soon learn to select those twigs that can easily be worked into the nest. The selection of a nest site, however, and the building activities themselves, are not changed by conditioning. Improvements in nest-building in older birds have often been interpreted as being caused by experience, but it seems to be settled beyond doubt that it is the consequence of maturation, first-year birds often building clumsy nests because of low (hormonal) motivation (Lorenz, 1937*b*).

This random selection of pertinent facts shows that in many species there are differences between various parts of the behaviour pattern, with regard to susceptibility to change by learning processes. As part

FIG. 110. Herring gull incubating on empty nest in full view of eggs.

of the innate equipment, such species have inherited certain strictly specific 'dispositions to learn'.

Preferential learning. The limitations of learning by innate endowment are not restricted to this phenomenon of localization, as the following instance will show. When the herring gulls' reactions to chicks and to eggs are compared more fully, it is found that, although youngsters about 5 days old are recognized individually wherever they are encountered within or even outside the territory, the reactions to the eggs are, as we have seen, not conditioned to the eggs themselves. They are, however, very rigidly conditioned to the locality of the nest. When a broody herring gull has to choose between the empty nest and the eggs in an artificial nest about a foot away, it will sit in the empty nest because it is on the accustomed site (Fig. 110). The bird will retrieve the eggs, or even occasionally sit on them in the artificial nest, but the attachment to the original site, acquired during the building of the nest, is extremely strong. When the eggs are removed a little farther from the nest, the gull will not respond to them by brooding but will eat them.

This shows that conditioning intervenes in reactions to the eggs as well as to the young, but that the bird selects very different qualities in the two cases: locality in the case of the eggs, neglecting the very conspicuous individual differences of the eggs; subtle individual characters in the case of the young. Here we discover another principle: it seems to be a property of the innate disposition that it directs the conditioning to special parts of the receptual field.

Similar examples have been studied and analysed in the case of orientation to landmarks of hive bees and digger wasps. These insects by no means condition themselves to every available landmark in the environment: they exercise a definite choice. The principles on which this choice is based are not identical in the various species, and are very different from the principles on which man would make a choice (Hertz, 1929, 1930, 1931; Tinbergen and Kruyt, 1938).

Critical periods of learning. There is not only localization of learning in relation to the reaction concerned, but there is also a certain localization in time, giving rise to the phenomenon of critical periods.

The Eskimo dogs of east Greenland live in packs of 5–10 individuals. The members of a pack defend their group territory against all other dogs. All dogs of an Eskimo settlement have an exact and detailed knowledge of the topography of the territories of other packs; they know where attacks from other packs must be feared. Immature dogs do not defend the territory. Moreover, they often roam through the whole settlement, very often trespassing into other territories, where they are promptly chased. In spite of these frequent attacks, during which they may be severely treated, they do not learn the territories' topography and for the observer their stupidity in this respect is amazing. While the young dogs are growing sexually mature, however, they begin to learn the other territories and within a week their trespassing adventures are over. In two male dogs the first copulation, the first defence of territory, and the first avoidance of strange territory all occurred within one week (Tinbergen, 1942).

Other, more or less similar phenomena are described by Lorenz (1935). Young geese (*Anser anser*) follow their parents soon after hatching. If the parents are removed, however, before the young have seen them or if the young are hatched in an incubator, the young attach themselves to another bird or even to a human being, should this be the first creature they meet. Once the young have adopted the wrong animals as their parents, it is impossible to make them accept members of their own species, even their own parents. This conditioning process takes no longer than a minute or even less. It has been called 'Imprinting' (*Prägung*) by Lorenz, who claims that it is irreversible. Although some authors doubt whether this process of imprinting is fundamentally different from other types of conditioning and believe that there are only differences of degree between them, it is clear that we have to do with an extreme restriction of learning ability to a critical period.

VII

THE ADAPTIVENESS OF BEHAVIOUR

INTRODUCTION

ALTHOUGH we are still far from understanding the mechanisms which maintain life in its multitude of forms, it is obvious that a living organism is in an extremely unstable state. One of the main characteristics of the living organism is that it possesses an overwhelmingly complicated system of mechanisms that protect it against adverse influences of the environment and enable it to maintain itself as a living organism. It is these mechanisms that the biologist studies, whether he is an anatomist, a physiologist, or an ethologist. Causal study of these mechanisms is not complete unless their contribution to the primary activity of life is demonstrated.

Many different terms pivot around the concept of self-maintenance. 'Biological significance', 'adaptiveness', 'directiveness', 'purposiveness', 'survival value', 'ecological function', &c., are all related concepts. They are all intended to indicate the fact that the mechanisms and/or structures considered contribute to the maintenance of the organism.

A peculiar narrowness of modern human thinking is indicated by the fact that relatively few biologists seem to be willing to give their attention to both the causes *and* the effects of observed life processes. Many of them confine themselves to the study of the causes *underlying* the observed phenomena, as do the majority of physiologists, while others prefer to study the way in which life processes contribute to the maintenance of life, as do most ecologists.

As has already been indicated, there is a tendency among biologists to regard the study of directiveness as incompatible with that of causation. In the physiologically minded worker this often leads to a certain neglect of the problems of directiveness, while the student of directiveness, as a reaction to this one-sidedness, tends to argue not only that the study of directiveness is just as important as that of underlying causes, but that it should have priority, or even that it should be undertaken *instead* of physiological study. As a typical representative of this last type E. S. Russell may be quoted. This author, in a recent work, points out that 'to treat them [organic activities] as purely physicochemical, to seek always for a causal explanation, and to analyse them without end, is to get bogged in a vast intricacy of unrelated detail, and to lose sight of the biological significance of these activities, their active relation to the life of an organism as a self-maintaining, reproducing and developing whole' (Russell, 1945, p. 4). 'Instead of ignoring the objective "purposiveness" of vital activities and assuming that somehow it can be accounted for by mechanistic processes of evolution, I shall put it in the foreground' (l.c., p. 8).

In my opinion, this is based on a double misunderstanding. First, it is certainly not right to identify 'causal explanation' with 'causal analysis'. Whereas mere analysis does indeed lead to 'a vast intricacy of unrelated detail', a causal explanation is based upon analysis accompanied by continuous re-synthesis; and such an explanation unites the details into a synthetic picture in which the details are not unrelated but arranged in a system.

The demand that directiveness should be studied *instead* of causation seems to me to be another misunderstanding. The two ways of studying life processes are not opposed to each other. The adaptiveness or directiveness of many life processes is a matter of fact and can be revealed by objective study; however, a description of the directiveness of life processes is not a solution of the problem of their causation. Once the survival value of a process has been recognized and clearly described, the biologist's next task is to find out how its mechanisms work; in other words, on what causal systems it is based.

Now although I want to argue the necessity of studying both causation and adaptiveness, I want to stress at the same time that it is of the utmost importance to distinguish clearly between the causal factors underlying the behaviour of the individual and the biological significance of behaviour. This is the more urgent since the theory of natural selection has shown that the 'biological end' served by an adaptive mechanism (of any kind: behavioural, or 'physiological' *sensu stricto*, or growth leading to the formation of 'adaptive' morphological structures) may act as a causal factor in directing evolution. This is the reason why the biological significance or the survival value of such mechanisms can in a sense be considered as causal factors too. When, however, the fundamental difference between these latter 'factors' (ultimate factors, Baker, 1932) and the causal factors at work in the mechanisms of the individual (proximate factors, Baker, 1932) is not clearly recognized, this terminology, derived from two widely different fields of zoology, may lead to a revival of the confusion that was prevalent in the period of old-fashioned teleology, when, in the study of the living animals' functions, actual causal factors and the biological end were indiscriminately considered to be effective ('proximate') causal factors.

In this chapter the adaptiveness of behaviour will be discussed according to this point of view.

The first problem to be discussed is: Is all behaviour adaptive, does every instinctive activity contribute to self-maintenance? This can only be settled after detailed study of concrete cases. As I hope to show, such a study usually reveals a high degree of adaptiveness. In fact, the more fully behaviour (or any other life process) is studied, the more evident it becomes that we have to do with highly 'improbable', intricately adapted means of maintenance.

On the other hand, it is no use denying that there are behaviour elements that may be non-adaptive in this sense, that are merely by-products of—in themselves adaptive—physiological processes.

Thus we have seen that the phenomenon of neurogenic causation of movement may lead to a situation in which there is a surplus of motivation. This surplus of motivation may lead to a vacuum activity or to a displacement activity. Optimal releasing stimulus situations acting upon a poorly motivated animal may give rise to intention movements. In all these cases the special movement might be entirely devoid of survival value; and as every movement acts as a sign stimulus on predators who are visual feeders, they might even be definitely harmful. The difference between non-adaptive and adaptive behaviour elements is most clearly demonstrated by a comparison of functionless displacement reactions or intention movements with such as have acquired a secondary function as a social releaser (see p. 171). Thus displacement 'fanning' in the male three-spined stickleback acts as an outlet of sexual motivation, but it has no meaning as a social releaser. Displacement digging acts as a threat to other males and thus serves to keep intruders out of the territory, which is a function with definite value (see p. 175).

There is another way in which adaptiveness fails to tell the whole story. Each form of behaviour may be adaptive and yet have characteristics that can only be explained historically. For instance, pigeons and sand-grouse drink in a different way (viz. by sucking) from other birds, which scoop water. These two modes of behaviour are both adaptive. The differences, however, are of historical origin; the two movements are not homologous but they are convergent solutions based on different foundations. In this way all behaviour is at the same time adaptive and non-adaptive, and both aspects are found by comparative study. The study of convergences helps to show up adaptiveness; the study of differences in convergent behaviour types helps to reveal the basic historic background.

Further, there are activities that are definitely adaptive in one respect while definitely disadvantageous in another respect. The following is an instance of this conflict between different types of survival mechanisms. Outside the breeding-season, the three-spined stickleback has a highly adaptive concealing colour-pattern. The back is much darker than the ventral side, thus counteracting the effect of dorsal illumination ('counter-shading'), and the sides show a pattern of vertical bars, thus breaking up the visual outline of the body ('disruptive pattern') (Fig. 111). During the short courtship period the melanophores are extremely contracted, leaving the back a radiant bluish-white, while the ventral side becomes a deep red. The result is a total reversal of the countershading while the disruptive coloration disappears (Fig. 112). The fish is now very conspicuous indeed, which is an adaptation to attract females. At the same time, such a male is extremely mobile and practically loses its escape reactions. There are indications that these adaptations serving co-operation between the sexes render him very vulnerable to predators like cormorants and herons.

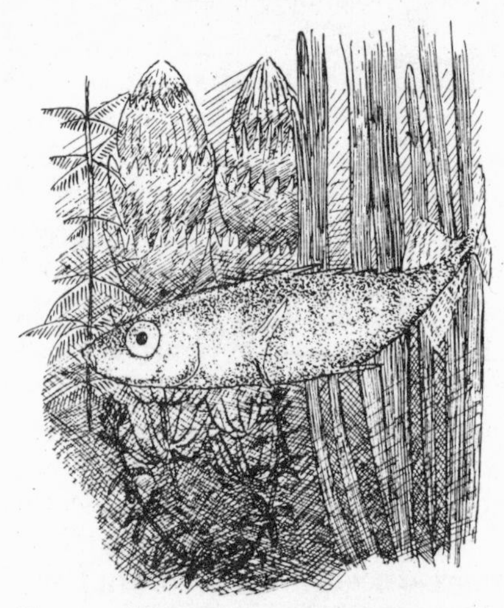

FIG. 111. Male three-spined stickleback in neutral, concealing coloration.

FIG. 112. Male three-spined stickleback in sexual coloration; the disruptive bars have disappeared, countershading is reversed.

Such a conflict between various 'interests', of which this is only a random example, is in fact a basic phenomenon of adaptiveness. The conflict is resolved by all animals in such a way as to compromise between the different demands. In the case of the conflict between conspicuousness and concealing coloration the compromise consists in the development of structures that confine the conspicuousness to the short periods during which it is actually needed. At other times the conspicuous colours are either covered by obliterative patterns, as in the case of birds, or they disappear by means of chromatophore reactions, as, for instance, in fish.

Another conflict is that between spring pugnacity and social attraction in gregarious species. In gulls and terns, for instance, territorial aggressiveness tends to push the pairs wide apart, while colonial life enables them to attack predators in force. The compromise is found in the establishment of rather large territories within a breeding colony.

A second problem is: To what degree is an animal, or one of its behaviour elements, adapted?

It is often stated that animals are 'perfectly adapted'. This is right only in a very meagre sense, meaning 'sufficiently adapted to maintain themselves'.

To cite a simple instance, the shrimp *Crangon crangon* has wonderful capacities for concealing itself from predators. Verwey gives the following account of its relations to *Sepia*:

> As a rule the animal lies hidden below the sand during the day and is busy during darkness. But much of its time is used in migration: to the shallower warmer water in spring and a good deal of summer, back to saltier water in autumn and part of the winter; it also moves up and down from the channels to the shallows, thereby following the tides. These movements are not restricted to the hours of darkness; they also take place during the day. Moreover, the animals creep out of the sand at the slightest inducement. All this causes a great restlessness so that their hiding is far from permanent.
>
> In this connexion it is right that the shrimp does not only possess its innate burrowing behaviour, but also the possibility for active colour change. It is not only dark on a dark background and light on a light one, but it assumes an appropriate tint for a yellowish or more brownish background by movements of appropriate types of pigment (Parker, 1930). This means that, when not on the move, the animal is not so easily seen.
>
> Enemies of the shrimp are among others the bullhead, the plaice, the dab, and the cuttlefish.
>
> . . . the one, acquainted with the shrimp probably best of all, is the cuttlefish. It makes its way just over the sand bottom, its head somewhat lower than its rear. Its funnel is directed on to the sand below and in front of it. It blows small water jets on to the sand. They hit a shrimp, which feels itself more or less exposed and begins to shove new sand over its back. I believe this is what the cuttlefish sees. Instantly it has seized its prey.
>
> The shrimp's hiding is perfect; its camouflage is perfect. But the animals preying on it are also perfect, each in its own way (Verwey, personal communication).

The expression 'perfect' has, of course, to be taken *cum grano salis*. It is clear that both predator and prey are, in a sense, imperfectly adapted, for the shrimp is actually caught very frequently, and the cuttlefish has trouble getting as many shrimps as it could eat and hence has to take other prey to a considerable extent. In reality, there is constant competition between the two, a neck-and-neck race, each species developing the adaptive means to meet the demands of maintenance; the struggle for life has led to the development of a highly specialized behaviour pattern with all its accessory instruments like eyes and effector organs in the cuttlefish, and to the development of concealing behaviour with all its accessories such as, for example, the concealing coloration, in the shrimp.

This struggle between predator and prey is especially interesting in the case of wasps (*Vespa* species). As described above, most songbirds learn to avoid wasps because of their unpalatability and their sting, which render them noxious, and of their conspicuous pattern which facilitates learning in the young birds during their first encounter with a wasp. The adaptations of the wasps could be called perfect. However, some species of birds, in spite of the wasps' defence methods, prey upon them regularly; among these are shrikes and flycatchers. This shows that the wasps' adaptations are not perfect and are 'open to improvement'.

However, these considerations apply to only part of a species' total adaptive equipment. Each species possesses a number of adaptations, and our next thesis is that these various adaptations supplement each other and may replace each other in different species. When a species *A* has less perfect powers of concealment from predators than a species *B*, it may yet maintain itself, for instance, by more rapid escape reactions or by greater resistance against, for example, some abiotic factor. Or a species may be inferior to another in resistance to adverse climate but superior in intelligence, as man is.

In discussing the survival value of behaviour it must also be borne in mind that behaviour is only one aspect of an adaptation, and hence cannot be separated from the other physiological and even anatomical features of the animal. Specialized feeding behaviour cannot be considered apart from the sense organs, nervous mechanisms, and effector organs involved in it; it is the working structure as a whole that has survival value.

What, then, is the part played by behaviour in this whole?

We have seen that the causal analysis of instinctive behaviour leads us to the conclusion that each animal has a number of instincts, each dependent on the activity of a nervous centre which controls, by means of a hierarchical system, a great complex of activities. We have seen that activation of an instinct causes seeking behaviour which, after a series of movements of increasing specificity, ends with the accomplishment of a consummatory act. Now the adaptiveness of behaviour is to be found in the fact, stated again and again, that this mechanism enables, and even forces, the animal to do 'the right thing at the right moment'. This implies that the mechanism forces the animal to withhold each reaction until the situation is reached which supplies the sign stimuli that call forth the very reaction that 'fits' the situation, i.e., that helps the animal to maintain itself. The 'directiveness' of behaviour therefore has the following aspects: (1) the selective sensitivity to specific situations, (2) the ability to perform directive movements, and (3) the highly specialized nervous mechanisms that are responsible for the fitting together of situation and response so as to give the response survival value. In short, it is the co-ordination of mechanism and objective purpose that brings about the adaptiveness of behaviour. The progress made by ethology or by biology in general is not achieved by the mere reiteration of this fact, for the fact is obvious right from the beginning, but the growing knowledge of how this adaptiveness is brought about.

Now, in spite of a fairly general acceptance of these views by students of behaviour, it seems to me that the concrete study of adaptiveness does not get as much attention as is desirable and that a short review of a limited number of examples of behavioural adaptiveness might therefore be of value.

Instinctive behaviour has usually been classified according to the various biological ends it serves. Feeding instincts, escape instincts, reproductive instincts, and so forth are just so classified. As was demonstrated in Chapter V, this classification runs parallel to a classification based on the underlying neurophysiological mechanisms. Hence instincts as classified in this chapter are the same as those classified in Chapter V. There is, however, another classification which cuts across the neurophysiological classification. While many behaviour elements are directly advantageous to the individual, there are other elements that are of no direct benefit for the individual but are of advantage to a group of individuals. In social species each instinct may contain such elements. In sexually reproducing species the reproductive instinct contains few elements that have survival value for the individual. Therefore one has to distinguish in each instinct between 'individual elements' and 'social elements'.

The social elements, viz. activities serving to maintain the species, possess an aspect that has not been considered until now. Individual development is based upon a system of growth processes of enormous complexity, a system that is different for every species. This is why interbreeding between species that are not very closely related usually leads to lethal disturbances in the development of the offspring. Therefore it is of survival value to the species to prevent interspecific mating. Every species, therefore, can only survive when it has developed sexual isolating mechanisms. Since, in sexual reproduction, union of the gametes depends more or less on active collaboration between male and female, behaviour plays a part in nearly every type of sexual isolating mechanism.

I will present the facts in the following arrangement: (i) Activities of direct advantage to the individual; (ii) Activities of advantage to the group.

Before proceeding to a review of examples so classified, two cases may be cited: the first illustrates the adaptiveness of a behaviour pattern as a whole, and the second shows the combination of behavioural, 'physiological', and anatomical features involved in 'an adaptation' (an adaptive character).

The fanning movement of a male stickleback ventilating the eggs consists of a combination of movements of the pectoral fins which directs

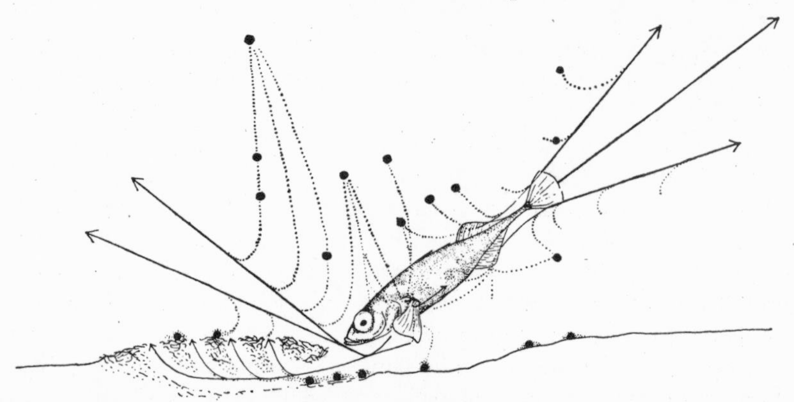

FIG. 113. Male three-spined stickleback ventilating the eggs. Dots indicate position of $KMnO_4$ crystals, dotted and solid lines indicate observed currents. By courtesy of I. Kristensen.

a current of water against the nest, and of compensatory swimming movements of the tail which keep the fish in the same place. The result is the production of a strong water current directed at the nest's entrance and so aerating the nest. By visual orienting reactions in relation to the nest, the nest entrance, and the surrounding vegetation, the fish manages to maintain a position in which the current ventilates the nest most effectively (Fig. 113). These orientation mechanisms have all been experimentally demonstrated.

The amount of fanning is regulated by both internal and external factors. Internal factors cause a gradual rise in total frequency in the course of the 7–10 days that are usually necessary for the development of the eggs. The total duration of fanning increases during this time, and so does the oxygen consumption of the eggs. Internal factors also shorten the intervals between separate successive bouts of fanning: these intervals may last for between 60 seconds and 60 minutes at the beginning of nest-care, but they decrease to approximately 5–20 seconds during the last days. In a similar way, the duration of each bout of fanning is gradually lengthened.

The influence of the external factors is superimposed on that of the internal factors. For instance, artificial aeration of the nest causes an appreciable drop in fanning activity; supplying the nest with water of low oxygen content increases fanning. A very low oxygen concentration releases another activity: the fish widens the opening of the nest. All these reactions and regulations can be studied physiologically, that is, with regard to their causation: at the same time their advantage to the offspring is clear. I should like to point out the fact that most of the experiments mentioned in this case would never have been carried out and would have made no sense at all if the recognition of the biological significance of the fanning movement had not come first. Why, for instance, should we try the influence of oxygen, rather than any other element or compound, if we were not guided consciously or unconsiously by the problem of finding out how an adaptation works?

The various aspects of an adaptation as a whole are well illustrated by the following instance (from Kepner, 1925, cited by and quoted from Lashley, 1938, pp. 445–6):

Microstoma, related to the more familiar *Planaria* and liver flukes, is equipped with nematocysts or stinging cells like those of the hydroids, which it discharges in defense and in capture of prey. In discharging, the stinging cell evaginates a threadlike barbed tube through which a poison is ejected. The striking fact about the creature is that it does not grow its own weapons, but captures them from another microscopic animal, the fresh-water polyp, *Hydra*. The *Hydras* are eaten and digested until their undischarged stinging cells lie free in the stomach of *Microstoma*. The nettles are then picked up by amoeboid processes of the cells lining the stomach and passed through the wall into the mesoderm. Here they are again picked up by wandering tissue cells and carried to the skin. The stinging cells are elliptical sacs with elastic walls, which are turned in at one end as a long coiled tube. In discharging, the wall of the sac contracts and forces out the barbed poison tube from one end of the sac. The nettle cells can therefore only fire in one direction. When the mesodermal cell carries the nettle to the surface, it turns it around so as to aim the poison tube outward. It then grows a trigger, and sets the apparatus to fire on appropriate stimulation.

When *Microstoma* has no stinging cells it captures and eats *Hydras* voraciously. When it gets a small supply of cells these are distributed uniformly over the surface of the body. As more cells are obtained they are interpolated at uniform intervals between those already present. When a certain concentration of cells is reached, the worm loses its appetite for *Hydras* and, in fact, will starve to death rather than eat any more of the polyps, which are apparently not a food but only a source of weapons.

Here, in the length of half a millimeter, are encompassed all of the major problems of dynamic psychology.

ACTIVITIES OF DIRECT ADVANTAGE TO THE INDIVIDUAL

Feeding Behaviour

The selective sensitivity of feeding behaviour is a problem encountered in every species. No animal eats everything, but each exerts a definite choice. Lack has shown recently that even closely related species may have different food, especially when they live in the same environment. Thus the foods of the shag and of the cormorant in Britain are different, though at first sight they seem to have very much the same hunting method. The shag eats more free-swimming forms, the cormorant specializes in bottom species like flat-fish, shrimps, and prawns (Lack, 1945).

Such differences may be due partly to differences in selectivity, partly to differences in the motor aspect of hunting. Thus a kestrel eats fewer birds than a hobby, partly because it is not able to catch adult songbirds on the wing; occasionally kestrels are seen to try to catch flying birds, thereby showing that the hunting instinct is more or less responsive to them, but such attempts rarely meet with success.

Feeding reactions of blood-sucking insects to mammalian hosts are released and guided by a number of stimuli. A comparison of, for instance, the fly *Stomoxys calcitrans* (Krijgsman, 1930) and the bug *Rhodnius prolixus* (Wigglesworth and Gillet, 1934) shows that although both react to a combination of sign stimuli each typical of mammals, they do not use the same stimuli. The approach towards the host is activated in *Stomoxys* by chemical, humidity, and temperature stimuli. Visual stimuli from the moving host play no part, nor do vibratory stimuli. In *Rhodnius* the latter two stimuli release the approach, and so do chemical and temperature stimuli; but moisture is without effect.

It is of special interest when one species shows different reactions to different food sources. Räber (1949) studied the stimuli necessary to release the predatory responses of the tawny owl. He found that small mammals are not recognized by their shape alone, for they must also move in a very special way: models were captured only if they had moving legs. Birds, however, were recognized by their shape: an immobile bird dummy was captured. According to Räber, these specializations are probably due to learning; it is striking how very well they are adapted to the natural situation. In the hours of twilight, when the tawny owls do their hunting, most birds are sleeping but mice are active.

Herring-gulls, like crows, some vultures, and probably other birds show a special reaction to food that is too hard to be crushed with the bill; they take it up into the air and drop it. An accident showed us that hardness is the property that releases the reaction in herring-gulls. In order to study the maturation of the incubation drive, I put a wooden egg on the territory of a gull before it had any eggs of its own. Until very shortly before the eggs were laid, the wooden egg did not release brooding behaviour. However, in some cases it released feeding responses, and once I observed that a gull, after giving a vigorous peck at the egg, at once took it in its bill, flew up in the characteristic way so commonly observed on the beach, and dropped the egg from a height of about 10 yards (Fig. 114). This was repeated twice.

The change of food preferences over a period of time is equally interesting. Noll studied the food preferences of young captive black-headed gulls and found regular fluctuations in food preference in the course of time. In early spring they preferred fish; from March till July they refused fish but took insects. From August till October fish was again preferred, but during winter fish were again left alone. These seasonal fluctuations in the captive birds, which were raised from the egg, coincided exactly with the seasonal fluctuations in habitat of the wild population from which they were taken. The individuals in the wild did not fish in the open water during the breeding-season and during actual winter, but in the autumn and the early spring they turned to fishing.

A similar fluctuation in food preference was found by Noll in the curlew. His captive curlews accepted earthworms only during summer and refused them at the onset of migration time. At this time wild curlews change their habitat, which coincides with a change of diet (Noll and Tobler, 1924).

FIG. 114. Herring-gull dropping wooden egg Explanation in the text

Such innate changes are of much more general occurrence than is usually recognized. Heinroth (1909) describes the female nightjar's tendency to pick up white objects shortly before egg-laying; this is also well known in the domestic fowl.

This phenomenon of 'specific appetite', of which the *Microstoma* case reported above is another fine instance, may also be much less rare than the scarcity of published records would make us believe, and my own experience leads me to believe that they occur in our own species more often than medical science recognizes.

It is scarcely necessary to point out the adaptiveness of the motor aspect of feeding behaviour. Yet some few instances may be cited. It is a curious fact that the most specialized cases have drawn so much attention that it is often forgotten that even the simplest method of feeding is already highly 'improbable' and in a sense specialized. This has led to a certain neglect of the relatively less specialized methods of feeding. Yet they are well worth studying, both from an ecological and an ethological point of view.

The larva of the water-beetle *Hydrous piceus* preys on snails of various species, preferably *Planorbis*. These are caught by pressing them between the folds of the back and thus handing them, by a combination of most improbable coiling movements, to the mandibles that finally crush the shell (Bols, 1935).

The European oystercatcher's method of opening mussels (*Mytilus edulis*) has been described by Dewar (1908, 1913). His description of the methods used to open mussel shells shows that, while the birds have command of a number of methods adapted to different positions of the mussel, the method most usually employed begins with an attack at the dorsal side. During falling or rising tide, when the mussels are partly covered by sea-water, they keep their shells slightly opened. The oystercatcher sinks its bill into the slit, vertically and in line with the valves. It then walks around, turning about its own bill as a pivot, and thus forces the valves apart. Often the bill is employed as a lever in another way, viz. after insertion of the bill, by lowering the head almost to the ground at one side of the mussel. The bill then presses on the opposite valve, thus opening the shells widely.

Escape Behaviour

The instinct of escape from predators may be taken as another example of behaviour beneficial to the individual. Receptor processes have already been mentioned in describing the recognition of birds of prey by other birds; here the releasing mechanisms are selectively receptive to the sign-stimulus 'short neck'. However, other sign stimuli are also important, and these may differ from one species to another. Thus geese respond to any object that is moving slowly. Speed is judged in relation to size of the object. A large aeroplane, a soaring sea-eagle, and a drifting downy feather all release escape responses. This releasing mechanism seems to be evoked in nature by the sea-eagle. Gallinaceous birds, which do not meet the sea-eagle, do not react to this sign stimulus.

Other species of birds seem to have different reactions to different birds of prey. Crows, for example, may crouch for a peregrine falcon but go up in the air for a goshawk under the same conditions. This is certainly adaptive, for a peregrine prefers to swoop down on a bird high up in the air, whereas a goshawk prefers to pick it up from the ground. It is not known whether this differential response is innate or not.

In addition, the different sorts of behaviour of a predator will elicit different responses from its prey. For instance, geese react to a slowly soaring eagle by warning and making preparations for escape, but a surprise attack from the eagle releases flight.

FIG. 115. Sand-dollars and starfish. Arrow indicates weak tidal current. Further explanation in the text.

In lower animals, selective responsiveness to special predators may be based on chemical sign stimuli. Thus both *Nassa*, a Gastropod mollusc, and *Pecten*, a Lamellibranch mollusc, react to the smell of a starfish by escape, which consists of a series of zigzag leaps in *Nassa* and of the spectacular swimming response in *Pecten* (Hoffmann, 1930; von Uexküll, 1921).

The Pacific echinoid *Dendrastes excentricus* (the sand-dollar) has a similar specific responsiveness to the scent of a starfish. When a starfish is thrown among a group of sand-dollars there is no immediate reaction, but in the course of a few minutes the animals in the immediate environment and those downstream dig themselves in (Fig. 115).

The adaptiveness not only of sensory perception but also of motor response is well worth a closer study. Chicks of gallinaceous birds react to predators either by crouching or by taking refuge in the 'danger shadow' of the mother. A similar (but not homologous) reaction is found in the young of several cichlid fishes, where the young crowd under the parent fish. In mouth-breeding cichlids the young, when not too old, hurry back to the mother's mouth (Peters, 1937; Baerends and Baerends, 1949). The greater spotted woodpecker, upon seeing a sparrow-hawk, hides behind or under a branch and keeps motionless. Birds and mammals of steppe and tundra show a very interesting convergence in that they run away from predators in a zigzag course (Fig. 116); this is known, for instance, in the European hare, in the nandu

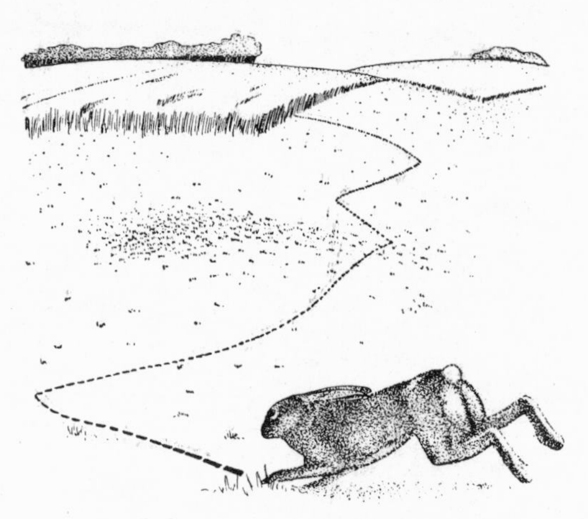

FIG. 116. Escape reaction of European hare.

(Portielje, 1925), and in ptarmigans (Krätzig, 1940), and undoubtedly occurs in other species as well.

Animals with specialized concealing coloration have equally specialized reactions to match the situation. Caterpillars with countershading, for instance, always take a position in which the darker side is turned upward. Species like *Dicranura vinula* and *Smerinthus ocellatus* that are dark on the ventral side turn this side upward. Species like *Gonepteryx rhamni* that have a dark dorsal side turn that side upward. The larva of *Apatura iris* has a dark head and is almost white at the caudal end; its normal posture is in accordance with this type of countershading (Fig. 117).

While most species of fish are countershaded, being dark on the back and white on the belly, there is a species that has a reversed countershading. This remarkable fish, *Synodontis batensoda*, is often seen to swim upside down (Fig. 118).

The position adopted by various insect larvae feeding on pine needles shows a similar correspondence with colouring. Seven species observed in Holland, having a pattern of longitudinal stripes of alternating green and white, all adopt positions with the needles parallel to their long axes

FIG. 117. Countershading and corresponding position in caterpillars. Above: *Dicranura vinula* in normal (right) and reversed position (left). Centre: *Smerinthus ocellatus* in normal (right) and in reversed position (left). Below: *Apatura iris* in normal position (right) and in uniform light (left).

(Fig. 119). In the hawk-moth *Hyloicus pinastri* the young larvae have this longitudinal pattern, but in the last instar the pattern is broken up into separate patches. Correspondingly, an individual of the last instar no longer alines itself parallel to the needles, but parallel to the twig. Furthermore, some of these species have a brown head of precisely the same colour as the brown leaves at the base of needles, whereas others

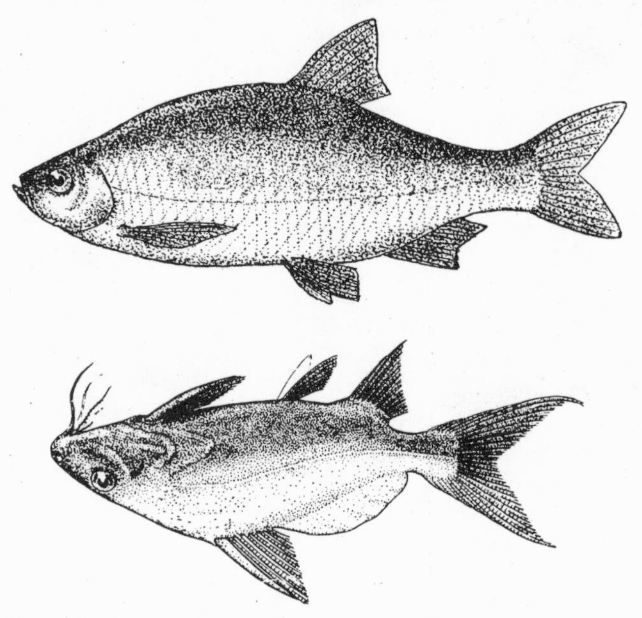

FIG. 118. Countershading and corresponding position in *Leuciscus* spec. (above) and *Synodontis batensoda* (below). After Norman, 1947.

have a green head. The former invariably adopt a position with the head between the brown leaves; the latter sit with their heads on the needles. The correspondence of behaviour and coloration, whether concealing or aposematic or coloration of any other type, is amazingly detailed. Owing to an over-critical attitude, a reaction to the over-elaborate 'arm-chair' teleology which preceded it, the survival value of such visual adaptations has been doubted and even denied. On the basis of recent experimental evidence, however (see, e.g., Sumner, 1934, 1935*a*, 1935*b*; Iseley, 1938; Dice, 1947), the survival value is now beyond doubt in a number of instances.

Adaptiveness of motor responses and of sensory responsiveness, especially the linkage of receptor and motor response in such a way that the production of adaptive responses in the appropriate situation is guaranteed, is based for the greater part on adaptive organization of the

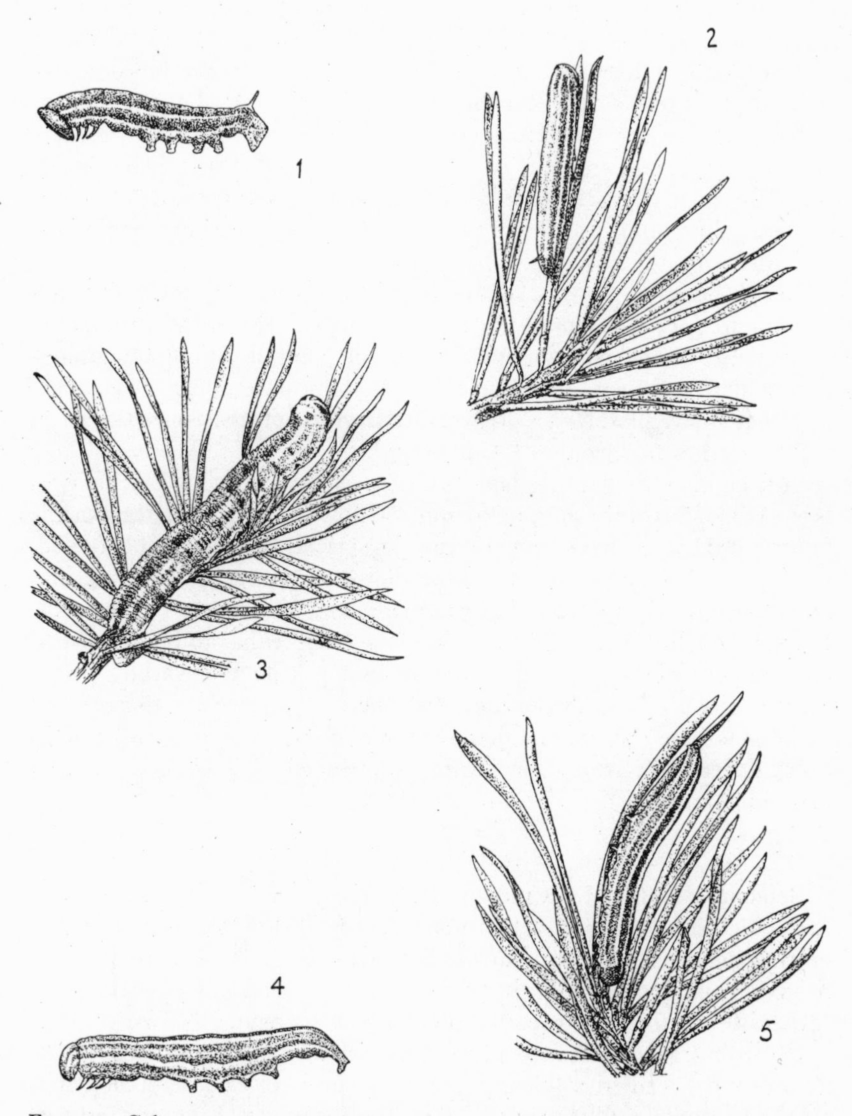

FIG. 119. Colour patterns of caterpillars living on pine needles. 1. Pine hawk-moth, fourth instar. 2. Same in natural position. 3. Same, last instar, in natural position. 4. *Panolis griseovariegata*. 5. Same in natural position.

central nervous system. A few instances might be mentioned now which emphasize the adaptiveness of central nervous organization in another way.

The peculiar 'localization' of learning processes in the reproductive behaviour pattern of the herring-gull was mentioned above: the birds condition themselves to their young independent of their exact position, whereas conditioning to the eggs is centred on the nest site and not on the individual characteristics of the eggs. The adaptiveness of this is evident: the young wander around over the large territory while eggs stay in the nest.

Honey-bees can be trained to confine their visits to special feeding-stations to special hours of the day. This is the reason why they (and presumably wasps too) so easily develop the habit of visiting the dining-room chiefly at lunch- or tea-time. The biological significance lies in the fact that many of their feeding-plants have a specific time-pattern of maximum honey secretion (Wahl, 1933).

Still another aspect of adaptiveness in central nervous mechanisms is the balancing of the automatic impulse flow to the centres responsible for the various behaviour patterns. As has already been mentioned, some patterns receive a much larger 'allotment' of the available impulse flow than others, and this rationing schedule is well adapted to the needs. Because, for instance, the sexual flight in *Eumenis* has to be performed many times a day, whereas 'playing the broken wing' in a warbler is only occasionally needed, the former gets such a large ration that central nervous fatigue can only be demonstrated after repeated stimulation in quick succession, while 'playing the broken wing' is easily exhausted experimentally.

ACTIVITIES OF ADVANTAGE TO THE GROUP

Behaviour that is of advantage to the group is more complicated than that which serves the individual alone. As a rule, it involves co-operation between at least two individuals. Although the adaptiveness of their behaviour is a property of the whole co-operative act, the behaviour of each of the individuals presents problems of its own.

Crowding, as such, may be advantageous in feeding. Many social birds, for instance, feed in a flock. The advantage is clear where the social attitude is combined, as it usually if not always is, with a strong tendency to sympathetic induction, as demonstrated by Katz and Révész in the domestic fowl.

In the instinct of escape from predators more cases of co-operation are known. Many species of social birds have evolved means of warning each other when a predator is observed. The conflict between individual security and security of the flock is very evident. Each utterance of an alarm call in small birds, for instance, is of disadvantage to the individual. As a consequence, solitary species have evolved no alarm note and no social reaction to the sparrow-hawk. Social titmice, on the contrary, give warning in reaction to the sparrow-hawk, and each individual not only utters the alarm call but also has a releasing mechanism responding to this alarm call which calls forth the escape responses: keeping silent and motionless in the cover close at hand (Kluyver, 1947).

FIG. 120. Flock of European starlings, undisturbed.

European starlings flying in a flock react to a peregrine falcon by drawing close together so as to form a very dense flock (Figs. 120, 121), and performing swift turns with a marvellous degree of co-ordination which is much more precisely performed than in the absence of a predator. A study of the hunting method of the peregrine falcon reveals the high degree of adaptedness of this response. The peregrine falcon is able to catch a flying bird by swooping down at enormous speed, exceeding 150 miles per hour. This high speed makes it extremely vulnerable to collisions. The only way in which it can collide with a solid subject (the prey) without hurting itself is by striking its prey with the strong talons first. This is the biological significance of the fact that a peregrine falcon takes care not to swoop down on and across a dense flock of starlings, but carries out a series of sham attacks until one or a few birds, by slightly inferior manœuvring, lose contact with the flock. It is only then that the real swoop is carried out. Incidentally, it is interesting to note that many different species have evolved the same defence method as the starlings. Ducks and teal and many waders behave in essentially the same way, and to the trained observer the sudden formation of numerous dense flocks above the mud flats, where just before multitudes of migrants were quietly feeding, is a sure indication of the presence of a hunting peregrine falcon.

Other species of birds carry out genuine attacks on a predator. These attacks are done in force, the mechanism of co-operation being as follows. An individual reacts to the predator by calling the alarm. The companions react to the alarm by joining the first bird, and attack together. The survival value of the perceptual process involved in spotting the predator lies in the fact that the discovery of the predator by the flock

FIG. 121. Reaction of starlings to peregrine falcon.

is dependent on the reactivity of the most wary or attentive individual. Also, with many pairs of eyes in action, the flock has a better chance of discovering a predator than the single individual. The survival value of the call-note and of the releasing mechanism responding to it is self-evident. The survival value of the social attack differs from one species to another. Sparrows rarely attack a cat, but collect above it in a bush continuously calling the alarm call. The survival value of this behaviour might be found in the advertising effect. Other birds, like terns and gulls, harass a predator to such a degree that his attention is seriously impaired and that even his actual movements are mechanically hindered; also, this behaviour probably conditions a predator to avoid the places where such attacks have repeatedly occurred (Lorenz, 1931). Jackdaws not only advertise the presence of a predator and attack it, but carry out specially furious attacks when the predator carries a jackdaw in its claws or mouth. Some species have two or more different social reactions to predators. Many birds, for instance, have two alarm calls, one responding to a surprise attack and another to a predator that is not yet attacking but is potentially dangerous. The reactions to these different alarm calls are entirely different; the first usually gives rise to crouching and similar emergency reactions, the second either to social attack or to seeking a hiding-place very carefully, as in most gallinaceous birds, or to preparatory measures of other kinds. It is very interesting indeed to study the adaptiveness of such behaviour by comparing the predator's behaviour with the social behaviour shown in response to it. Everyone who seriously attempts such a study will be surprised by the refined adaptiveness of every detail.

Thus we see that in social co-operation of benefit to a community an extra link, that of intercourse between individuals, is inserted between the initiatory perceptual process and the ultimate adaptive response. This extra link is a kind of signal system. The signal is given by one individual and it releases the response in one or a number of other individuals. In the cases already mentioned, the alarm call releases the social attack, &c. While the problem of adaptiveness in the perceptual processes of the initiating individual and that of the motor response in the reacting individual or group are not essentially different from the same phenomena of individual behaviour, except that they serve the group instead of the individual, the 'inserted link' is, in a sense, a new feature, and deserves special attention.

Since Lorenz started a new experimental analysis of these phenomena a number of facts have been brought to light which show that social co-operation depends on the delivery of a signal, which may be a special movement, the demonstration of special structures, the emanation of a scent, a special call, &c. These signals have been termed *Auslöser*, which has been translated 'releasers'. In order to avoid confusion (which may easily arise from the fact that a releaser is not anything that releases a reaction) I will use the term 'social releaser'. A social releaser, therefore, is a device especially adapted to release a response in individuals of the same species. The releasing mechanisms of the responses shown by the reacting individual seem to be innate in most cases.

The following paragraphs will demonstrate the various ways in which this concept of social releaser is realized in the organization of animal communities. I will confine myself to the reproductive activities because here social releasers attain their highest degree of specialization.

In the reproduction of the higher animals the following elements may be involved: mating, sexual fighting, and care of the offspring.

Mating

The survival value of mating itself is so evident that it need not be discussed.

The first step in mating is the attraction of a mate. This can be effected by a reaction to the breeding-site in both sexes, as with many birds that return to the same breeding-site year after year, or, in addition, with the help of a social releaser, by which one sex attracts the other. A device in use in groups with good sound receptors (vertebrates, some insects) is a loud call-note. Some of the 'mating calls' in male birds have the function of attracting the females. Song in most songbirds, the 'love call' of a heron, drumming in woodpeckers, 'bleating' in snipe, rattling of nightjars—all these are social releasers the (partial) function of which is the attraction of females from afar. The same function is fulfilled by the chirping of locusts and crickets and by many calls of frogs and toads. Strangely enough, the experimental proof of this function is still very meagre. Proof has been given only for the chirping of crickets and locusts. Regen (1914) had male *Gryllus campestris* chirp before a telephone receiver. The sounds were transmitted to a loudspeaker stationed in another room where mature females were kept. The females approached the loudspeaker. In a field test, Duym and Van Oyen put mature females of the locustid *Ephippiger ephippiger* at equal distances from two concealed boxes, in one of which were kept normal males that chirped regularly, while the other contained males whose stridulating mechanism was silenced. The females invariably went to the singing males (Fig. 31, p. 35) (Duym and Van Oyen, 1948).

In birds few experiments have been performed so far, although the existence of excellent gramophone records, especially in America, makes it possible for every field-worker to carry out the most exciting tests.

Eygenraam, by imitating the drumming of the great spotted woodpecker, released hostile responses (drumming in reply) from territory-holding males of this species (personal communication).

'Natural experiments' where birds reacted to the song of concealed individuals have been witnessed by many observers.

A comparative study of song reveals that (1) song is characteristic of each species, and (2) the females' response is very selective. This renders song a most important factor in reproductive isolating mechanisms. As we shall see, isolating mechanisms usually (always?) contain releasers and the corresponding reactions. In many species, copulation is even based on a series of social releasers, which together render interbreeding highly improbable and, as a rule, even entirely impossible.

Attraction by scent has been observed in, for example, various Lepidoptera. In most species it is the female that attracts the male, as in *Saturnia* species and in *Lasiocampidae*, where the females have special scent glands and the males have highly specialized antennae in which the chemoreceptors are located. In psychids the female is quite unable to fly; the male is attracted by scent and mates with the female right on the puparium (Matthes, 1948).

Visual attraction plays a role in many animals with well-developed eyes. In birds a ceremonial flight adds to the effect of song in a number of species, but sound is more important than sight in almost every case. In fish it is often the visual stimulus that plays the most important part.

The nuptial colours of the three-spined stickleback attract the females. The ten-spined stickleback has a pitch-black colour in spring. Just as in the three-spined stickleback, the influence of this nuptial coloration of the male on the female has been proved by experiments with dummies (Sevenster, 1949). This striking difference in nuptial coloration between two closely related species again shows the specificity of such devices.

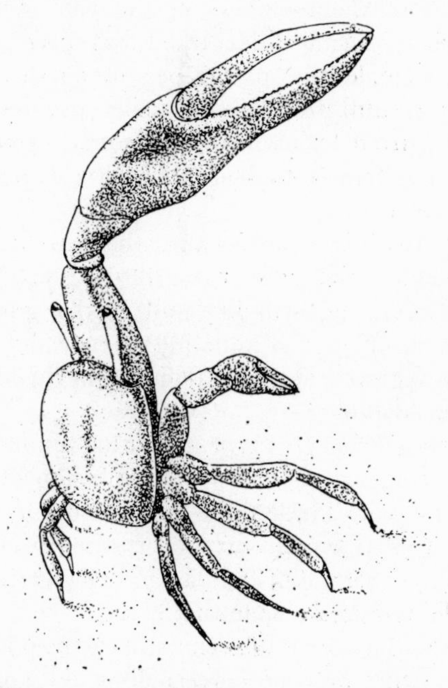

FIG. 122. Male fiddler-crab (*Uca pugilator*) waving its claw. After Pearse, 1914, from Verwey, 1930.

In fiddler-crabs, attraction of females seems to be dependent exclusively on manner of movement. The manner of waving of the large claw (Fig. 122), which serves to warn off males as well as to attract females, is very specific and it is amazing to see how it is possible to create such striking differences within one such simple type of movement. Crane, having studied the behaviour of twenty-seven species of *Uca* at Panama, concludes that 'each species proved to have a definite individual display, differing so markedly from that of every other species observed, that closely related species could be recognized at a distance merely by the form of the display. Furthermore, related species had fundamental similarities of display in common, and series of species, showing progressive specialization of structure, in general showed similar progression in display' (Crane, 1941, pp. 152–3).

After the male has attracted a female, copulation may occur at once or an intervening period of 'bethrothal' may be inserted. During this period nuptial activities may be performed frequently and with high intensity. The biological significance of this phenomenon has not yet been extensively studied, but some observations render it probable that they effect a synchronization as well as a general stimulation of the reproductive states of the members of the pair. The synchronizing function of 'display' during this period has been demonstrated in pigeons by Craig. Female pigeons can be stimulated to lay eggs by the administration of stimuli that are normally provided by the male. Ovulation can be started by giving the adequate tactile stimuli alone, or by permitting the female to see and hear a courting male from a distance (Craig, 1911).

Fraser Darling (1938) has put forward the hypothesis that many of the display activities and the more or less diffuse reproductive behaviour going on in colonies of sea-birds like gulls, terns, guillemots, gannets, fulmars, &c., have the effect of stimulating the reproductive mechanisms. Although there is scattered evidence pointing in this direction, definite proof is still not available.

It should be stressed that this type of mutual influence is only one of three factors that bring about the synchronization of sexual activities in the members of a pair. These three factors work in succession.

As a first step, a gross synchronization is often brought about by an external rhythm (lengthening of the day and changes in temperature in most species of the temperate zone that breed in spring).

As a second step, individual differences in the responsiveness to these general external factors may be corrected by the general stimulating function, in Darling's sense, of social releasers synchronizing the maturation processes in the sex glands.

A third step is the exact timing to the minute and even second. This is the task of the majority of sexual releasers. Where the mechanism of sexual co-operation has been relatively well studied, the co-operation between male and female depends on a series of releaser-effects. This has already been mentioned and analysed in the case of the three-spined stickleback (p. 49), of *Helix pomatia* (p. 50), and of *Eumenis semele* (p. 32). As will be clear, the part played by releasers here, as in attraction, is not only the timing of the response but also its spatial orientation. These examples need be elaborated no farther. It is necessary to emphasize the general aspects, viz. (1) the extreme degree of adaptive specialization attained in these releaser relationships, and (2) their function as isolating mechanisms. The latter aspect is especially obvious when all the sign stimuli necessary for any given animal to accomplish mating are summed up. For the female stickleback, a mating partner must be red and must perform a zigzag dance; it must lead her to the nest, show her the nest entrance, and deliver the very special tactile stimuli which the male delivers while 'quivering' (p. 48). Though each one of these sign stimuli may not be sufficiently specific and 'improbable' to serve as an isolating mechanism, together they certainly are improbable and specific enough to prevent fertilization by any other species.

The principle of releasers was discovered gradually. While the function of call-notes and song has been realized early, although dimly, that of visual releasers is not even generally recognized today (see Rand, 1941, 1942; Tinbergen, 1948). The theory of visual releasers throws an interesting and clarifying light on many cases of brilliant 'nuptial' coloration. As we shall see, the function of these conspicuous colours is not merely that of social releasers in the service of actual copulation, but it is certain that part of its function is to bring males and females to well-timed and well-oriented co-operation.

As was mentioned above, every animal that is vulnerable to predation by visual feeders develops protecting devices, mostly some type of concealing coloration. In species where mating releasers develop as well there is a conflict between these two functions: releasers have to be conspicuous, while escape behaviour tends to bring about inconspicuousness. The result is that the display of releasers is usually confined to a minimum. This minimum may apparently be very high; for instance, one might think that many songbirds could do with a little less persistent song. However, one should not be too rash in judging these things. On the one hand, what knowledge we have of the action of selection makes it evident that predator pressure tends to eliminate excessive and really harmful display of social releasers. On the other hand, it is just in these cases that intraspecific selection might cause a species to break through the moderating influence of natural selection so that it runs 'out of control'. We will return to this problem in the next chapter.

Fighting

Fighting is, in the great majority of cases, between individuals of the same species. Moreover, fighting is usually subordinated to the reproductive instinct. Most fights are between individuals of the same sex. An animal that is attacked by a predator may fight, but only relatively few species have developed this type of fighting. While in the case of fighting against predators the survival value is of course evident, the question whether fighting between individuals of the same species has survival value might give rise to doubt; yet it definitely has.

Reproductive fighting, while potentially fatal to certain individuals, is in general advantageous to the species as a whole. Since the publication of Howard's *Territory in Bird Life* (1920) a great number of papers has been published in which instances of sexual fighting were described and their function discussed. It has become clear that the survival value of sexual fighting is to be found in the fact that it divides certain objects, which are indispensable for reproduction, among as many males as possible. These objects are different for different species, as the following examples will show.

In many species of birds and other animals, fighting results primarily in the acquisition of a territory. The result of spring fighting in, for example, songbirds, or lapwings, or sticklebacks is the staking out of territories, in each of which there is a male who, by the intensity of his attacks, keeps trespassers off.

It has been doubted whether the territory system really sets a limit to the number of males in a given stretch of country. Occasional observations of successful intrusions by newly arrived males, resulting in a decrease in size of the original territories, have been interpreted as proving that fighting does not prevent the establishing of new territories by any males that might seriously attempt to do so. However, for every male that succeeds are numerous males that do not. Also, observations in captivity and in the field show that the defence of a territory-owner becomes more vigorous when his territory is getting smaller. Huxley has indicated this by saying that territories are like rubber disks (1934). The result is that although, theoretically, there is no minimum size, there is, practically, a minimum size below which defence becomes so vigorous as to frighten off all intruders. In captivity this may even lead to a situation where one male kills all other males within the same cage.

A territory may be of advantage in various respects. In most songbirds, territory is necessary for the provision of a minimum quantity of food in the immediate vicinity of the nest, so that, even under adverse climatic conditions, the collection of food during the first days after hatching does not take too much time. As the young stop begging when they get too cold, shortness of foraging trips is an absolute necessity (Tinbergen, 1939*a*). In other cases, as in kestrels and many hole-nesting species, territory is primarily a matter of a suitable nesting-site (L. Tinbergen, 1935).

In other cases, such as ruffs and blackcocks, territory protects from intrusion and disturbance of coition by other individuals (Lack, 1939).

Another function of sexual fighting is the acquisition of other objects indispensable for successful reproduction. In the bitterling (*Rhodeus amarus*), a fish which lays its eggs in the mantle cavity of freshwater Lamellibranchs, usually *Anodonta*, sexual fighting is restricted to the vicinity of the mussel. When the latter moves about, the territory moves with it. In the carrion-beetle (*Necrophorus*) fighting is centred around the carrion (Pukowski, 1933).

The survival value of most sexual fighting, if not all, is to be found in the acquisition of a sex partner as well. This is especially clear when the territorial factor is entirely or almost absent, as in the moose, in deer, and in other mammals. In some birds, too, 'free' fighting occurs (great crested grebe, Huxley, 1914; the European avocet, Makkink, 1936). However, in territorial species, also, a certain amount of fighting occurs outside the territory. This can be seen when the female wanders beyond the boundaries. Another indication of the importance of the sex partner is the fact that even in some of the most extreme territorial species males fight males and females fight females (Tinbergen, 1936*a*). Sexual fighting, therefore, serves to acquire a territory, a mate, a mussel, or other objects indispensable for reproduction.

The survival value of acquiring a sex partner is, of course, primarily the guaranteeing of coition itself, that is, of fertilization, with its two aspects of coming into contact with a sex partner and preventing disturbance by other individuals. However, as Heinroth pointed out, fighting of males against males and of females against females, as found in monogamous species, is a means of guaranteeing the successful rearing of the young.

Now it is a very striking and important fact that 'fighting' in animals usually consists of threatening or bluff. Considering the fact that sexual fighting takes such an enormous amount of the time of so many species, it is certainly astonishing that real fighting, in the sense of a physical struggle, is so seldom observed. This is due to the fact that fighting has its disadvantages as well as its advantages as a means of maintenance. It is, as we have seen, of advantage to keep individuals well spaced out; however, it is a distinct disadvantage when individuals are actually damaged or killed and thus excluded from reproduction. The compromise that has developed is to have releasers that intimidate without causing damage. This is why sexual fighting is often accompanied by an elaborate display of 'gladiatorial vestments'.

Releasers in the service of sexual fighting have a double function. An individual within its territory or in the neighbourhood of a female is stimulated to attack any individual displaying the characters of a rival. An individual outside its territory or far from its female is stimulated by the same display to withdraw without fighting. When the fighting-drive in the two individuals is about equal, such as in territorial boundary disputes, lengthy threat performances appear in both of them long before actual fighting occurs.

The same instinctive inhibition of the fighting-drive is found in man. One reason why wholesale slaughter in modern warfare is so relatively easily accomplished is to be found in the modern long-range arms that prevent one witnessing the action of lethal weapons. Our instinctive reluctance to kill is strengthened by the sight of a dying man in mutilated condition. Hence one is much less reluctant to direct artillery fire at a distant tower, thereby killing the enemy artillery observer, than to cut his throat in a man-to-man fight. Our instinctive disposition has not changed with the rapid development of mechanical long-range killing apparatus.

Some instances may be cited of the effect of releasers in connexion with sexual fighting.

The song of male songbirds, apart from attracting females from afar, serves at the same time to repel other males. A very clear 'natural experiment' in which both the intimidating and the fight-provoking function are demonstrated is mentioned by Lack in the English robin.

> On May 27th, 1937, an unringed newcomer, evidently wandering without territory, started to sing in a corner of the territory owned by a long-established resident male. The latter, then in a distant part of its territory, promptly sang in reply. The newcomer, which could not, of course, yet know that it was trespassing, sang again. The owner, having flown rather closer in the interval, sang again in reply. The newcomer again sang, the owner again approached and replied, now more vigorously, and this procedure was repeated twice more, the owner finally uttering a violent song phrase from only some fifteen yards away, but still hidden from sight by thick bushes. At this point the newcomer fled, from an opponent it never saw, nor did it appear again. (Lack, 1943, p. 29.)

As in the sexual domain, social releasers in the service of fighting may act on any kind of sense organ. The visual releasers may consist of a type of movement or posture, or a special, usually conspicuously coloured structure, or (most often) a combination of the two, that is to say, a movement specially adapted to display a conspicuous colour.

A typical threat-movement is found in the herring-gull. A male herring-gull attacking an intruder stretches its neck and at the same time points its head downwards (Fig. 66); although this posture may seem to be relatively unspecific to the casual observer, it is, for those who know the ways of herring-gulls, a most conspicuous and improbable posture, and it is full of dreadful meaning for other herring-gulls (Tinbergen, 1936*b*).

An extreme case of a releaser consisting of a gaudily coloured structure without a special display movement is presented by the cere of shell-parakeets (Fig. 123). Males of this species have a clear blue cere, females a dull brown one. As long ago as 1926, Cinat Tomson showed by simple but entirely convincing experiments that males judge the sex of newcomers by the colour of their ceres. Females with ceres painted blue were treated like males, that is to say, they released the whole pattern of reactions directed to males. Males with the cere painted brown were courted. When the animals become individually acquainted with each other this simple relationship is complicated by conditioning; a male cannot be induced to attack its own mate by painting her cere blue.

FIG. 123. Head of shell-parakeet.

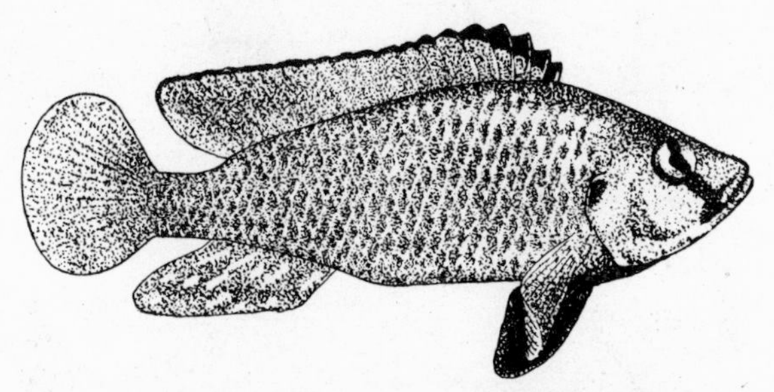

FIG. 124. Male of *Astatotilapia strigigena* in reproductive markings. After Seitz, 1940.

Fighting in the male of the cichlid fish *Astatotilapia strigigena* depends on stimulation by three visual releasers. A dummy must be (1) laterally compressed, offering a large surface to the male, (2) glossy silver or blue, and (3) spotted with 'jewels'. While (2) and (3) are purely morphological releasers (Fig. 124), (1) depends on behaviour. Not only does a male in fighting mood turn its broad side to the opponent, but it enlarges its surface by raising all its fins, a reaction that, in this type of fish, nearly doubles the surface displayed. Also, the fins are covered with jewels as well as the body surface, and erection of the fins at the same time intensifies the jewels' display.

Similar co-operation between behaviour and conspicuously coloured morphological structures is found in a large number of cases (Fig. 125). While the close correlation between the movements and the structures involved has been observed in many of these, experimental proof of their releaser function has been given in only a few species. In lizards the belly, which is displayed by lateral compression of the whole body, often has a bright colour in the males. In the European species *Lacerta*

FIG. 125. Some instances of correlation between display movements and conspicuous colour pattern.

1. Threat display of red breast in European robin. After Lack, 1943.
2. Sexual display of undertail covers in lapwing.
3. Threat display of 'epaulets' in chaffinch. After L. Tinbergen.
4. Sexual display of side in golden pheasant.
5. Threat display of gill-cover 'eyes' in *Cichlasoma meeki*.
6. Threat display of gill-cover 'eyes' in *Hemichromis bimaculatus*.
7. Male pintails displaying the back of their necks to female. After Lorenz, 1941.
8. Sexual display of back in snow-bunting. After Tinbergen, 1939.

viridis the male's blue throat has exactly the same function as the stickleback's red belly. A female with the throat painted blue is attacked by males, and also attracts females. A male with its throat painted green (the female's throat colour) is not attacked even by the most aggressive male. In the American species *Sceloporus undulatus* (Fig. 126) a female with the ventral surface painted blue (the male's colour) is attacked by males, but males painted grey are courted (Noble and Bradley, 1933).

In *Lacerta viridis*, as in the three-spined stickleback, it has thus been shown that the male's colour display acts upon both males and females. Although in most other cases only one of the two functions has been studied, it is probable that in these, too, the 'nuptial colours' have this double function.

A striking instance of convergence is offered by the cuttlefish *Sepia officinalis*, which has been studied by L. Tinbergen (1939). Parallel with

FIG. 126. Male *Sceloporus undulatus* displaying blue ventral surface. After Jaques in Noble, 1934.

the development of their eyes (convergent to those of fishes), the courtship of Cephalopoda has evolved into a typical visual one, closely resembling the courtship of certain sexually dimorphic fish. A male *Sepia* in mating condition assumes a strongly variegated pattern of alternating white and dark purple bars, and displays the most conspicuous part, the broad, flattened, lateral surface of the fourth arm, towards other individuals (Fig. 127). Reactions of males and females to this display differ essentially: a male returns the display, a female in mating condition keeps quiet and allows the male to copulate. Experiments showed that the male's nuptial colours, and especially the colour and display of the arm, released fighting in other males, while all models coloured and 'behaving' like females were treated like females.

This short review will be sufficient to show the wide diversity of social releasers serving to direct the fighting instinct to the conspecific rival and, on the other hand, to keep the fighting urge down to the level of threat, thus preventing undue damage by actual fighting. Although, in general, conspicuous structures which act as threat often serve the function of attracting females as well, there are differences between species in the relative importance of each of these two functions. In some species threat may be the primary or even the only function, in others the emphasis may lie on the release of mating responses. As in the case of social releasers serving to ensure sexual co-operation, the conspicuousness of threat releasers may cause a conflict with the require-

FIG. 127. Male *Sepia officinalis* in sexual display. After Tinbergen, 1939.

ments of the escape instinct, and this presumably is the cause of the fact that threat display is only shown when actually needed.

Care of Offspring

Releaser relations between parents and offspring have been studied in relatively few cases. On the receptive side we know (p. 149) that selective reactivity to the eggs or young may be dependent either on innate releasing mechanisms alone or on IRMs on which conditioning is superimposed. Lorenz relates that pheasants and ducks may incubate the eggs of other species, but that they kill the chicks when they hatch. Whether this selectivity is innate or conditioned as in the cichlid fish studied by Noble (p. 53) is not known. Lorenz thinks it may be an innate reaction to the colour pattern of the head.

The gaudily coloured inside of the mouth in nestling passerine birds displays every character of a releaser, being very simple, very conspicuous, and very specific—perhaps another example of an isolating mechanism serving to confine the parental behaviour to the offspring. It would be interesting to study this problem in connexion with parasitic birds that have acquired similarly coloured mouth patterns.

The releasing value of begging movements and calls of young birds is also well worth investigating. In general those young that display the most intensive begging movements are fed first by the parents. Many species also have special warning signals, like the well-known alarm calls in birds. Some cichlid fish, e.g. *Cichlasoma biocellatum*, 'call' the young and stimulate them to gather in the 'danger shadow' by special movements like shaking of the front of the body (an intention movement).

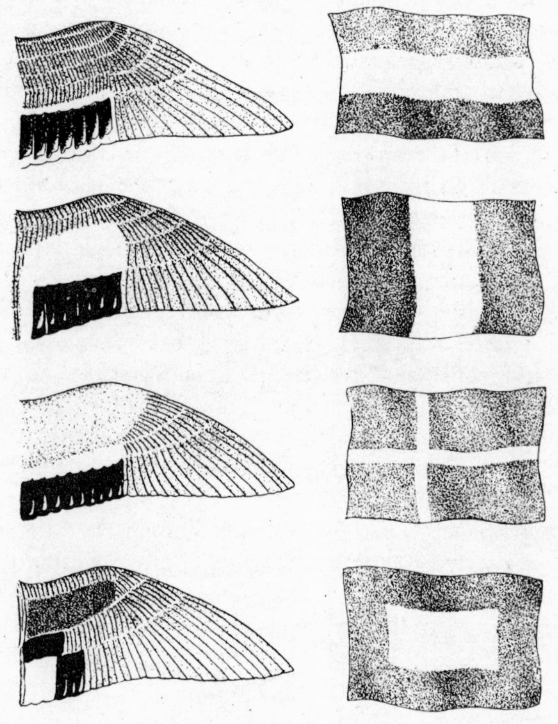

FIG. 128. Wings with specula in various ducks, and flags.

The behaviour involved in care of the offspring shows the same refinement of adaptiveness as behaviour of other types. However, apart from birds, most species have been studied rather superficially, and even in birds we are still far from having a precise knowledge. To mention an instance of specialized feeding behaviour: it has recently been observed that the great tit has a very special way of feeding large caterpillars to the small young. It tears the head off, then feeds the young one by one by squeezing the contents out gradually, feeding one or two drops to each young, much as one squeezes toothpaste out of a tube (Kluyver, 1948).

Summarizing this paragraph on social releasers, it will be clear that although their function has been experimentally proven in relatively few cases, we can safely conclude that they are adaptations serving to promote co-operation between the individuals of a conspecific community for the benefit of the group. It is a striking fact that all social releasers studied seem to be beautifully adapted to activating an IRM, for a social releaser is always specialized in such a way as to send out stimuli that have the characteristics of sign stimuli. They are always relatively simple and at the same time conspicuous. This is why Lorenz considers the social releasers as adaptations to the IRM.

Further, social releasers show specialization in the direction of specificity ('improbability'). This is not only a means of guaranteeing the release of specific responses, but it serves to confine the influence of social releasers to individuals of the same species. Since the great majority of social releasers are developed in the service of reproduction, their specificity causes them to act as reproductive isolation mechanisms.

It is very interesting to notice how these requirements of simplicity, conspicuousness, and specificity have been compromised by selection in exactly the same way as in signal structures developed by man by purposeful and conscious designing. This has been emphasized by Lorenz, who pointed out the striking similarity of visual releasers, like the wing-specula of ducks, and national flags (Fig. 128).

In conclusion I may emphasize once more the fact that this chapter is not intended to be an exhaustive treatise on the adaptiveness of instinctive behaviour. It merely serves to demonstrate the necessity as well as the possibility of an objective study of directiveness of behaviour. This demonstration is a natural consequence of the thesis that behaviour can, and should, be studied in the same way as any other group of life processes.

VIII

THE EVOLUTION OF BEHAVIOUR

INTRODUCTION

ETHOLOGY, as well as biology in general, is centred around three major problems, viz. (1) the problem of causation, (2) the problem of adaptiveness, and (3) the problem of evolution. Whereas this book is primarily concerned with the first problem, the two other problems could not be left out of consideration altogether. This final chapter will therefore be devoted to a consideration of the last major problem, that of evolution.

The study of the evolutionary aspects of behaviour has lagged far behind that of the morphological study of evolution. This is not surprising; the study of behaviour is still in its infancy in comparison to morphology. When we realize the enormous amount of labour, first of palaeontologists, then of comparative anatomists and embryologists, finally of geneticists, which has been necessary in order to arrive at even a sketchy and tentative picture of evolution, it must be clear that we cannot possibly expect to get a fair insight into the evolution of behaviour until behaviour has been much more extensively studied. Yet a better insight into the evolution of behaviour would be of paramount importance for our understanding of human conduct, and especially of human sociology.

Let us first consider the problem of why our knowledge of the evolution of behaviour is still so backward, for this will help us to prepare the way for well-planned study. In the first place, our lack of knowledge is due to the fact that the study of behaviour is much more elaborate than morphological study. Hence it takes incomparably more time to compose a behaviour monograph than a morphological monograph. Morphology can be studied from preserved material, but the student of behaviour depends upon his animal showing its complete behaviour in his presence, which in many cases may be a matter of years. Further, the description of behaviour, when actually observed, is much more difficult than the description of form. This is why we still possess almost no complete behaviour monographs. A second cause of the relatively backward position of ethology is the fact that it requires scrupulous study to trace innate elements in behaviour and to distinguish these components from acquired traits. In morphology the errors arising from confusion of phenotype and genotype are at present negligible. In ethology many behaviour elements have to be studied in both normal individuals and individuals raised in isolation, and the many conditioning processes taking place even in such isolated animals have to be followed closely. The work done thus far in pursuit of this type of knowledge is still very fragmentary indeed; we are badly in need of 'ethograms'—monographs of the entire behaviour pattern of separate species. Until now the most valuable work of this kind has been done with birds. The studies of Whitman, of Heinroth, and of Lorenz especially, who worked with hand-raised birds, are of great value.

Although these two factors responsible for our lack of knowledge can be countered by an intensification of ethological research, there are other intrinsic restrictions whose removal is not within our power.

Palaeontology

In morphology our knowledge of the lines along which recent animals have developed from their ancestors is based for the greater part on palaeontology. It is through numerous well-preserved fossils that we know, for instance, that the class of mammals has evolved from reptilian ancestors. In the South African Karroo formation numerous fossils have been found of a great number of species whose structure is intermediate between reptiles and mammals, and although the exact taxonomic position of each of these species is still open to some doubt, there cannot be the slightest doubt that together they represent reptilian groups that were on their way to becoming mammals.

While palaeontology is thus of enormous importance for our knowledge of the history of life as far as structure is concerned, providing us with genuine historical documents, it is practically useless as a means of understanding the evolution of behaviour. For fossils do not behave, and we will never know how long-extinct animals behaved. Of course we know something: for instance, as the extinct dodo and the solitaire had spurred wings, we may safely assume that these birds, like their living relatives the pigeons, used their wings in fighting. From an analysis of fossil tracks Abel (1921) and others have been able to tell us something about the locomotion of extinct species. But all evidence of this kind will always be both relatively uncertain and even more fragmentary than our knowledge of living forms. Palaeontology therefore, which has been such a valuable aid to the student of evolution of form, fails entirely as a source of documents about the behaviour of animals of the past.

Embryology

In morphology another source of data is embryology. Embryology, or the study of individual development, reveals to us the relatedness of many forms that are quite different in the adult stages. Through embryology we learn, for instance, that all the vertebrates, however different they may be in their adult stages, evolve from an embryo that is very much alike in all the classes. For instance, while the land vertebrates have a quite different arterial system from that of the fishes, both groups pass through an embryonic stage in which the arterial systems are very much the same, and the differences develop by divergence in later stages.

To cite one out of numerous instances: recent birds differ from most other vertebrates in that they have a very short tail skeleton. In embryos the tail is much more like that of other vertebrate embryos and consists of a number of separate vertebrae. Later, these elements fuse to form one bone.

Again, embryology has been of great help in understanding relationships within the group of crustacea. The systematic position of barnacles and cirripedia is at once made clear by their typical free-swimming crustacean larvae.

Thus we see that related forms show divergent development starting from very similar stages. Although Haeckel's formulation of this fact in his 'biogenetic law' (ontogeny is a recapitulation of phylogeny) has been shown to be over-simplified, the facts upon which it is based are of great importance, and embryology is still one of the great accessory sciences of evolution study.

One of the main sources of error in applying Haeckel's formulation is the frequent occurrence of 'caenogenetic' characters, that is, of recent adaptive acquisitions, in the larvae. Thus, though the long tail of a tadpole is certainly an indication that frogs and toads have evolved from ancestors that had long tails, the fact that some of them have horny teeth certainly does not mean that the ancestors of the Anura had horny teeth; they are a recent adaptation. For a complete discussion of these problems I must refer the reader to De Beer (1940).

Embryology, like palaeontology, is of little use to the student of the evolution of behaviour, and it seems doubtful whether it ever will be. The main reason for this seems to be the fact that 'caenogenetic' characters are even more common in behaviour than they are in morphology. However, there are some indications of recapitulation in Haeckel's sense. For instance, songbirds (Passeres) of most species hop instead of walk as most other birds do. Larks and pipits, however, walk. Now in these species the young pass through a hopping stage (Lorenz, 1937*a*). Although similar facts are known, the study of ontogeny of behaviour cannot be expected to contribute much to the study of evolution.

Thus the ethological facts at our disposal for a descriptive picture of evolution are relatively few in comparison with morphological facts. Indeed, as both palaeontology and embryology are of so little help, the only possible approach is through typology, based on comparative study of present-day, full-grown animals.

Typology

Typology in morphology is based upon the establishment of homologies. Even when considered apart from embryology and phylogeny, typological study allows us to homologize organs of separate species, that is, to recognize them as different developments from the same ancestral element. Thus, however different a bird's wing and a dog's front leg may be superficially, a close study of their anatomy shows that they both have, for instance, a humerus, an ulna, a radius. Similar, though less striking, similarities can be traced in the hand.

Or, to take another example, a comparative study of various crayfish and their relatives shows that they all have five pairs of walking legs built according to one general plan. In some genera, like *Homarus*, the first pair is very different from the other pairs in that it bears the huge claw. Now by comparison of a number of species one can find intermediaries, forming a chain of links between species like *Homarus* and those species where the first pair is not very different from the subsequent pairs.

Synthesis of Palaeontology, Embryology, and Typology

Now typology, while in itself leading to a definitely static picture of *Baupläne* or general structural schemes, around which the various concrete realizations are grouped, can be used as part of the dynamic picture of animal life provided by the study of evolution. This is done by interpreting the typological system as the result of divergent evolution, just as ontogeny of recent forms may be dynamically interpreted as the outcome of a historical sequence of ontogenies, and just as fossils may be interpreted as stages of a gradually changing animal world. The most important conclusion this synthesis of various independent sciences has led to is that their respective descriptive pictures of the history of animal life agree, thus mutually supporting each other. For instance, the conclusion reached by palaeontologists, that both mammals and birds must have developed as divergences from a reptilian stock, is confirmed by both ontogenetic and typological study. This coincidence of conclusions of course considerably strengthens the separate conclusions reached by each one of these sciences.

Once this is realized it will be clear that we are seriously handicapped in any attempt to study the course of evolution of behaviour. Not only do palaeontology and embryology fail us as independent sources, leaving us merely typology, but this means much more than reducing to one-third the weight of evidence at our disposal. It means that we cannot check results reached by comparative study with data from other fields.

However, there is another possibility of checking interpretations based on comparative behaviour studies. Each animal species has of course developed as a whole, and therefore any evolutionary interpretation of behaviour must fit in with interpretations of the evolution of the same group of animals reached through morphological study. In other words, any 'family tree' of a group based on ethological data has to be identical with one based on morphological data. If there is no conformity, either one or the other interpretation is wrong, or both.

The Dynamics of Evolution

Alone with descriptive work, the causes of evolution have to be studied. As in morphology, the main contribution to this causal study has to come from genetics. Here again the backward position of ethology is striking. Owing to the difficulty of tracing genetically determined behaviour components, geneticists have nearly always used morphological characters as indicators of gene function.

For this reason this chapter will be highly fragmentary and will make an even more urgent plea than hitherto for more intensive study.

FACTS BEARING ON THE DESCRIPTIVE STUDY OF THE EVOLUTION OF BEHAVIOUR

We shall now proceed to discuss the ways in which comparative ethology can give us an idea of how behaviour has evolved. As in comparative anatomy, the first thing to do is to get a clear idea about what should be compared.

The Establishment of Homologies

First, of course, we have to confine ourselves to the innate basis of behaviour. But this is not enough; comparison, in order to be of value, cannot be haphazard but has to use elements of the same order of complexity. In comparative anatomy it is evidently impossible to homologize organ systems as a whole, such as the arterial system or the skull. The arterial systems of two species may each be composed of a number of elements, some of which can be recognized as homologous, while others clearly are not. Thus homologization of the systems as wholes is not possible, while it is possible to homologize separate components. Thus it is found that the elements compared should not be too comprehensive. On the other hand, it is also possible to choose too small elements for the purpose of homologization: it clearly makes little sense to try to compare, for instance, one special cell in the liver of one species with one similar cell in that of another species. The choice of the integrative level of the morphological units to be compared evidently is of great importance. The morphological structure of an organism is a hierarchical structure. Cells are grouped together in cell groups, these into tissues; several tissues may together make an organ; these again may be integrated into an organ system. Homologization in comparative anatomy is practised at the intermediate levels, that of 'organs' and parts of them.

When applying homologization to behaviour elements we meet with a similar necessity of choosing the level of comparison. We have seen that innate behaviour, or the causal systems underlying it, are organized into a hierarchy. Simple reflexes and automatisms are integrated into relatively simple rhythms and patterns. These patterns in their turn are parts of complexes of a higher order, such as locomotion. Of still greater complexity are the innately fixed parts of consummatory acts, the 'fixed patterns'. These again function as component parts in instinctive, 'directive' actions. At which level shall we begin our attempts at homologization?

As in morphology, this is largely a matter of trial and error. Some instances might be given to illustrate the procedure. Let us compare the nest-building of various birds in order to settle the problem of whether the constituent activities are homologous or are the result of convergent evolution. What little we know at present shows that many species, even those not closely related, have much in common. They build in a sitting position, depositing the material around them and shaping the cup by frequent scraping with the legs. But there are numerous differences between the species. Let us take, quite arbitrarily, two species which I happen to know from personal experience: the European sparrow-hawk and the long-tailed tit. The sparrow-hawk, like many other birds, works the twigs into the nest-wall by peculiar quivering movements. This quivering is carried on until the twig sticks, and it thus serves to make a solid meshwork. This movement is not seen in such perfection in nest-building long-tailed tits. This species uses mosses, which are deposited on the nest's rim. Now and then a long-tailed tit collects a beakful of spider's threads, and with peculiar movements of head and neck weaves them all over the rim of the nest cup, thus providing solidity by 'sewing' the component mosses together. In both species the result of the building behaviour as a whole is very similar: a relatively strong cup. We will not enter into the many dissimilarities between the two species' nests, because they are irrelevant here. The main point should be clear: the two species both shape the cup of the nest by scraping. Solidity of the nest's wall, however, is brought about by different means: the sparrow-hawk makes quivering movements which are only weakly developed in the tit; the tit makes weaving movements which are not observed in the sparrow-hawk. Such a state of affairs leads us to the conclusion that we cannot simply say that nest-building is homologous in the two species. But it is possible to homologize parts of it, viz. the scraping movements. As these are found in many birds, for instance in (probably) the majority of Passeres, in the Laro-Limicolae, in birds of prey, &c., it is quite probable that homologization is justified. Weaving and quivering, on the contrary, are in all probability two entirely different elements.

Now it is quite possible that these concrete conclusions are not absolutely unshakeable. We simply do not have the necessary observational facts. Whether right or wrong, however, they suffice to demonstrate my point and thus to show that homologization is most profitable when practised at the level of behaviour elements like scraping, quivering, &c.

Now this is a significant fact. Purely descriptive study of the above type leads us to conclude that behaviour elements of the type of scraping and quivering are those that can be most easily homologized. It is remarkable that these are exactly the elements which physiological analysis has led us to consider as genetically determined, intrinsically co-ordinated units, the 'fixed patterns'. Although comparative study is still in its infancy, the scattered data we possess all point to the conclusion that homologization will have the best chances when fixed patterns are singled out as the units to be homologized. They play the same part as 'organs' in comparative anatomy.

It is interesting to note how several observers who, in their time, could have but a dim knowledge of the physiological basis of 'fixed patterns' have yet picked out just these elements as the units to be compared. Whitman, who as early as 1898 wrote, 'Instinct and structure are to be studied from the common standpoint of phyletic descent', gave us a monograph of pigeon behaviour (1919) in which he focused his attention especially on fixed patterns like 'billing', 'cooing', &c. Another pioneer, Heinroth, wrote a thorough treatise on the behaviour of Anatidae (1911) in which he used behaviour elements, mostly fixed patterns, as a basis for classification along with morphological characters.

Homologization of Social Releasers

As a special example of evolutionary study of behaviour elements based entirely on the method of comparison I choose the study of social releasers, because they are examples of relatively rapid evolution.

Many movements serving as social releasers appear to have taken their origin as displacement reactions (Tinbergen, 1939*b*, 1940; Kortlandt, 1940*a*; Lorenz, 1941). Such displacement reactions, originally serving as an outlet of a surplus of motor impulses, may, when secondarily acquiring survival value as social releasers, become 'ritualized', that is, adaptively refashioned according to the needs of a social releaser: simplicity, conspicuousness, and specificity (p. 184).

As an example, let us take the displacement fanning movements of a male three-spined stickleback. The reproductive cycle of the male stickleback consists of several successive stages. Of these, only the sexual stage *sensu stricto* and the subsequent egg-stage are of interest here. Having fertilized a certain number of clutches, a male loses its readiness to court a female and to fertilize eggs. Instead most of its time is now taken up by ventilating the eggs. This 'fanning' movement can easily be distinguished from other behaviour.

Now this fanning movement of the male is used as a displacement reaction in situations where there is a surplus of sexual motivation (see p. 116). When a female arrives that is unwilling to spawn, the male, being sexually aroused, will vent the sexual impulses by bouts of displacement fanning. This fanning has no influence on the female; it does not serve as a social releaser. It is also scarcely (if at all) different from genuine autochthonous fanning.

As described on p. 49, the courtship of the male consists of a series of releasers, serving to evoke the female's sexual reactions: the zigzag dance, leading, showing the nest entrance, and quivering or trembling. Close study of the movements by which the male shows the entrance revealed that it resembled fanning more than any other movement. As the movement occurs when the male is strongly motivated sexually although it cannot yet perform the consummatory act (sperm-ejaculation), it was concluded that 'showing-the-entrance' is displacement fanning. The difference between this type of displacement fanning and autochthonous fanning can be understood as visual ritualization, that is, adaptation to the function of visual releaser tending to make the movement conspicuous. The female actually reacts to visual properties of this movement.

Furthermore, trembling also resembles fanning, and the same arguments can be applied in favour of considering trembling to be a type of displacement fanning. Trembling is also ritualized, but now as an adaptation to giving a mechanical stimulus. We know that the female reacts to the touch stimulus given by the trembling male.

These hypotheses were based on a close study of one species. After they were formed, a related species, the ten-spined stickleback (*Pungitius pungitius*) was studied and compared with *P. aculeatus*. It was found that the *P. pungitius* male shows the nest entrance to a female by a movement very closely resembling its particular mode of fanning. Thus comparison strengthened the conclusion that 'showing-the-nest-entrance' in both species is displacement fanning (Van Iersel, personal communication).

In this way comparison may help us to identify homologous displacement reactions in the species of one group and in tracing the evolutionary process of ritualization.

Among cranes, the fighting drive, when obstructed, finds an outlet in displacement preening. Lorenz (1935) has shown that this movement, which has a social releasing (threat) function, differs very much from one species to another. Whereas in some species it is but little ritualized and hence can be recognized easily as preening, it is in other species entirely different from preening. Comparative study, however, shows that there are still other species which are intermediate in this respect, and by studying these various intermediate species the threat movements of species like the Manchurian crane are recognized as extremely ritualized preening movements.

Although relatively many displacement activities acting as social releasers are already known, extensive comparative studies are still rare. The most complete study is that by Lorenz, concerning the behaviour of the surface-feeding ducks (Lorenz, 1941). Yet the scattered evidence we have at present is sufficient to show that displacement activities offer exceptionally good opportunities for the study of the evolution of adaptive behaviour elements.

Comparison of Larger Units

The fact that the development of morphology is so far in advance of ethology renders it possible to use comparison for the purpose of evolutionary study without a detailed knowledge of behaviour homologies. Once morphology has succeeded in giving us a picture of the phylogenetic relationships within a group, we may take this picture for granted, and compare the behaviour of the species concerned. Or, starting with behaviour, we might arrive at a tentative picture of the evolution of a certain type of behaviour and check this picture with the conclusions of morphological study. Thus it is possible to arrive at conclusions about the general trends of evolution within large groups without having made a detailed typological behaviour study. I will give some examples.

Friedman (1929) has attempted to trace the origin of the parasitic cowbirds. By comparative study he finds that all the relatives of cowbirds, and even some of the cowbirds themselves, are not parasitic and thus concludes that parasitism is a relatively recently acquired type of behaviour. This is a very general conclusion indeed, bordering upon a truism. More specific conclusions are at the same time more tentative. It is, however, interesting to note that the species showing many primitive morphological and geographical traits (*Agelaoides badius*) has also progressed less than others towards parasitism, in that it uses other birds' nests but incubates and rears its own young.

Davis (1942) has studied the reproductive behaviour of the four species of the Crotophaginae, a subfamily of the cuckoos. These birds show 'communal nesting', that is, they all live in flocks and build one nest in which several females lay their eggs. Of these species, anatomical studies show *Guira guira* to have most primitive traits; *Crotophaga major* is more specialized; and *C. ani* and *C. sulcirostris* are extremes. The breeding flock of *Guira* consists of pairs which may defend individual territories within the flock's territory, and sometimes one pair builds a nest of its own within the flock's territory. The flock of *C. major* is composed of pairs which unite to defend a territory. In *C. ani* the communal nesting has reached its climax. Polygamy or promiscuity is the general rule and the whole colony defends the territory. Thus those species that are considered primitive on morphological grounds also have the least specialized breeding-habits, and so we may safely assume that the sequence *Guira–C. major–C. ani* and *C. sulcirostris* represents, more or less accurately, three steps in the evolution of this type of nesting.

In a similar way the evolution of social behaviour in social Hymenoptera can be understood in rough outline. Among the ants there are no solitary species at all, but the majority of bees and wasps are solitary. There are numerous species in which the female guards the eggs and even the young. In some species the first generation of young even remain in the nest with the founder, and thus there are numerous species more or less intermediate between the solitary and the highest social condition. This points to the conclusion that the most highly developed states have evolved from families of mother and offspring (Maidl, 1934). In the termites the origin almost certainly is a family of male and female and their offspring.

Conclusions of a still more general kind can be drawn from a study of general trends in the evolution of major taxonomic groups. Thus, for instance, the general trend in a mammalian behaviour might be characterized by pointing out the relative paucity of fixed motor patterns and the development of 'plastic' behaviour. In comparison, insects have specialized in stereotypy of motor responses and in very specialized and stereotyped learning processes. Comparative studies of this type have never grown beyond more or less vague indications, yet consistent work along these lines could lead to a considerable deepening of our insight.

In the limited scope of this book I must content myself with this sketchy outline and refrain from touching on other sides of this fascinating field. I hope it has been shown clearly enough that a close comparative study of behaviour for the purpose of evolutionary interpretation is both necessary and possible.

Just as in morphology, however, this work is but a first step. It can never provide more than a preparation of the stage for investigations of the causal relationships underlying evolution. A few words should be said on this topic.

THE MECHANISM OF THE EVOLUTION OF BEHAVIOUR

In the study of the mechanisms underlying evolution, morphology is again far more advanced than ethology. Here the backward position of ethology is even more striking, for a genetics of behaviour still has to be developed. What evidence we have is incidental and fragmentary, which is only natural in view of the great difficulties met with in the singling out and describing of ethological 'characters'.

The natural thing to do under present conditions seems to be to follow the trails cleared by geneticists and students of evolution in other fields and to test the applicability of their theories to our field of study.

This, of course, should be done with a mind open to the possibility that an independent study of the evolution of behaviour might lead us to new viewpoints.

Present-day theories of evolution consider mutations in the widest sense as the basis of all heritable change. The variability due to mutational change may show directiveness of various types, adaptive as well as non-adaptive. Adaptiveness is brought about by selection. Speciation, or the divergent evolution of populations originally belonging to one species, starts with geographical expansion of the species' range to such a degree that two or more populations of one species become reproductively isolated. The various populations thus isolated are usually slightly different in genetical make-up right from the beginning. This difference, together with the environmental differences leading to different selection pressure, account for divergent evolution of the populations which ultimately results, via the formation of geographical races, in the origin of new species, genera, and even families. Whether this 'micro-evolutionary' process is at the bottom of all evolutionary divergence, even of those often called macro-evolutionary, is a matter of disagreement. It is certain, however, that the causes of evolution can only be studied in micro-evolutionary processes.

This being the case, various types of approach are possible. First, we should know whether and in what manner mutations do affect behaviour. The most direct way to settle this question would be by genetic experiments of the classical type. A roundabout way would be to study the behavioural aspects of speciation; that is to say, to study the behaviour differences between subspecies belonging to one species and between closely related species. Further, one should know whether behaviour characters have selective value. Apart from these questions there are reasons to consider whatever viewpoints might be suggested by facts known thus far. In ethology the problems of sympatric speciation (that is, speciation without geographical separation of populations) and of Lamarckian aspects of evolution have been raised.

Mutations and Behaviour

The experimental evidence on behavioural effects of mutation is scarce. Whatever evidence we have, however, points to the conclusion that innate behaviour elements are no different from 'morphological' or 'physiological' characters in their dependence on genetic constitution.

As an example of an experiment specially designed to study the genetical background of a behaviour character, Herter's study (1941) of temperature preferences may be mentioned. Herter, having determined the temperature preference of a white and a grey strain of domestic mice, crossed the strains. The F1 hybrids had exactly the same preference as the white parental strain, but in the second generation 12 individuals behaved like the white form and 7 behaved like the grey form. Back-crossing with white gave individuals with the same preference as white, while back-crossing with grey resulted in 16 animals with the same preference as white and 12 individuals behaving like the grey form. This behavioural character therefore segregated like a morphological character; in this case the temperature preference of white was dominant, and the colour itself was wholly independent of the behavioural character.

Facts of this kind represent the first step towards a genetics of behaviour.

Speciation and Behaviour

Whereas evolutionary change is generally far too slow a process to be open to direct study, there is one aspect in which evolutionary change can, so to speak, be seen in projection in one temporal plane. Evolution is not only a matter of change in each species in the course of time but also entails a splitting up of one species into two or more populations, each of which may in a relatively short time develop into separate species. This process of speciation is still taking place; and this is the reason why, by comparing cases where the splitting up has scarcely begun with such as have advanced farther, we can get a fair insight into the lines along which this speciation process is taking place. This is why so much of our knowledge about evolution is based on zoogeographical combined with taxonomic study.

By basing our study on the recognition of this fact and making a comparative ethological study of subspecies and closely related species, we shall be able to see the evolution of behaviour in the making. Of course, this again is still a dream, for this study has scarcely begun.

However, the results obtained thus far are very encouraging and a short sketch is worth while.

Subspecific differences. Relatively many instances are known of behavioural differences between subspecies of one species. These differences may concern the motor aspect of behaviour, as with bird-calls, or, more often, they may concern the receptive side of behaviour.

For instance, two subspecies may have a different habitat preference. The mistle thrush, *Turdus viscivorus*, may be an example (Peitzmeyer, 1942; Mayr, 1942). One population, which lives in France and Belgium, and which has recently spread north and east into Holland and across the Rhine into Westphalia, lives in deciduous forests. Another population lives in the mountainous regions south and south-east of Westphalia, and this population lives in coniferous forests. It is quite probable that this difference in habitat preference is innate.

Entirely comparable are the subspecific differences in temperature preference found in various animals such as *Carabus nemoralis* and *Lacerta sicula* (Herter, 1941). Such differences in Innate Releasing Mechanisms, of major instincts or of parts of them, seem to be among the first things affected by mutations.

The herring gull (*Larus argentatus*) and the lesser black-backed gull (*L. fuscus*) in north-western Europe are considered to be extremely diverged geographical races of one species, which, having developed by geographical isolation, have come into contact again by expansion of their ranges (Mayr, 1940). The two forms show many differences in behaviour; *L. fuscus* is a definite migrant, travelling to south-western Europe in autumn, whereas *L. argentatus* is a much more resident habit. *L. fuscus* is much more a bird of the open sea than *L. argentatus*. The breeding-seasons are different. One behaviour difference is specially interesting. Both forms have two alarm calls, one expressing alarm of relatively low intensity, the other indicative of extreme alarm. *L. argentatus* gives the high-intensity alarm call much more rarely than *L. fuscus*. The result is that most disturbances are reacted to differently by the two forms. When a human intruder enters a mixed colony, the herring gulls will almost always utter the low-intensity call, while *L. fuscus* utters the high-intensity call. This difference, based upon a shift of degree in the threshold of the alarm calls, gives the impression of a qualitative difference in the alarm calls of the two forms, such as might well lead to the total disappearance of one call in one species, of the other in the second species, and thus result in a qualitative difference in the motor-equipment. Apart from this difference in threshold, there is a difference in the pitch of each call.

Specific differences. Comparative study of a number of closely related species gives us an insight into further advanced stages of speciation.

Such studies have been made in various groups. Heinroth (1920), Lorenz (1941), and Delacour and Mayr (1945) have published accounts on the Anatidae (geese and ducks). As a result, this family is by far the best known. Whitman's work with pigeons (1919) is also of considerable importance, and recently Spieth (1947) has started a similar study in *Drosophila* which is a most promising subject indeed. These studies, and a number of less comprehensive ones, have made possible a first, tentative conclusion. Unlike subspecific differences, specific differences are not at all confined to changes in the IRM or to relatively slight differences in motor responses. Closely related species may differ considerably in their motor patterns.

Lorenz has pointed out that these differences in motor equipment seem to be found primarily in one special group of behaviour elements: the social releasers. Whereas, as we have seen, the 'fixed patterns' behave, taxonomically, as 'conservative' characters (that is to say, are not subject to rapid evolutionary change and are therefore useful as taxonomic characters), the social releasers, many of which are of the same order of complexity as fixed patterns are not at all conservative. This seems to hold specially for such social releasers as play a part in reproductive behaviour, perhaps because these are just the behaviour elements that act as isolating mechanisms, where there is, so to speak, a premium on specificity. On p. 193 I have already mentioned displacement preening as a threat releaser in cranes, and displacement preening as a releaser playing a part in the courtship of ducks. The ducks, being the best-known group, enable us to compare the behaviour differences with the correlated morphological differences. In every species studied the 'courtship preening' movement is made more conspicuous by the fact that a gaudily coloured structure is displayed by it. In the mallard, and in most other species, this is the metallic speculum of the wing. However, the colour pattern of this speculum is different in each species, and more so than the movement. In some species (*Lampronessa sponsa*, *Aix galericulata*) the movement is slightly different in that the bill touches one special feather which has a very conspicuous colour and shape. In the garganey teal (*Querquedula querquedula*) the bill is not moved along the inside of the wing but along the outside, and it points to a bluish-grey field of wing coverts. Throughout the group, differences in the movement are relatively slight, but differences in the conspicuous structures taking part in the movement are considerable. Morphological evolutionary changes therefore are more easily detected than changes in the motor pattern, at least in this case.

The evolutionary interpretation of these facts must be as follows. As displacement preening occurs in all the ducks studied by Lorenz and in none of the more distantly related geese and brent geese, these movements have to be considered homologous within the Anatinae, just as autochthonous preening itself. Their divergent evolution must be due to selection, as specificity is a demand imposed by their function as isolating mechanisms.

Spieth's study of *Drosophila* courtship (1947) provides us with similar facts. The most conspicuous posturing movement of the male is some form of wing vibration. This releaser is different in the various species studied. In *D. fumipennis* it is 'wing-waving', in *D. nebulosa* it is 'scissoring', and in *D. capricornis*, *D. sucinea*, *D. willistoni* and *D. equinoxialis* it is 'wing-vibration'. These three movements are no doubt divergent homologous forms of one genetically determined behaviour element.

The manifold forms of waving the claw in fiddler crabs are another instance (see p. 173).

When this phenomenon of divergent evolution, based upon rapid evolutionary change of behaviour components of the fixed pattern type, is carried to the extreme, it may lead to total disappearance of some of these elements. Only in this way can we understand why, within a group of closely related species, some species 'have' a certain movement, while in others their homologue cannot be found. Lorenz's paper contains many examples, and although negative evidence of this kind has a touch of uncertainty because further study might reveal homologies where they are not expected now, I think the dropping out of elements of behaviour is a reality.

Also, homologous behaviour elements may, so to speak, shift their position within the pattern in that they may come to serve different functions in different species. Thus, just as, in the course of the evolution of mammals from the reptilian stock, the quadrate and the articular, originally forming the jaw-hinge, have shifted their function to become incorporated into the mechanism of the ear, so a behaviour element like *Aufstossen* in the ducks (Lorenz, 1941) has functionally replaced the call note in some species, such as the pintails and the teals.

Sympatric speciation. While there is considerable evidence pointing to allopatric speciation (Mayr, 1942), that is, speciation through expansion and subsequent geographical and reproductive isolation, some authors claim that sympatric speciation may also occur. We must examine this hypothesis because in its most recent form (Thorpe, 1945) it is founded on ethological evidence. The advocates of sympatric speciation hold that reproductive isolation may become established between members of one population by a sudden change in habitat preference in certain individuals, which makes them colonize a habitat within the original range of the population, until then avoided. Coinciding with this change in habitat a change in mating preference would be established, preventing the new colonizers from mating with the conservative individuals. The colonizers' offspring would be genotypically attached to the new habitat or else would be conditioned to it. If the two habitats have a mosaic distribution, a splitting up of one species into two 'ecological races' would be the result. The newly formed 'race' would, by its different habitat preference, be able to colonize areas where the original species would not be able to settle. Thus ecological (sub) speciation would precede and lead to geographical (sub)speciation.

The evidence that led to this hypothesis is indirect. One of the main supports is the evidence of changed habitat selection in the adult *Drosophila melanogaster* as a consequence of conditioning in the larvae (Thorpe, 1939).

As Mayr (1948) points out, the hypothesis as a whole contains several unproved assumptions. First, the conditioning experiments, carried out by Thorpe and others, did not give results in all individuals; there certainly was a shift in habitat preference after conditioning, but a relatively high percentage of the adults did not choose the new environment even after the conditioning procedure. This indicates that in nature a considerable part of the colonizers' offspring (viz. that part which, although it had grown up in the new environment, would still not prefer it to the old) would return to the conservative population, thus tending to mix the two populations. Another difficulty is that it has not been proved that colonizers with a strong preference for the new habitat would not mate with the conservative individuals at all. Thus, although the ethological evidence is strong enough to lead one to consider the possibility of sympatric (ecological) speciation, it is still much too incomplete.

McDougall's Work on Lamarckism

There is another way in which ethological work has led to views on the dynamics of evolution which are not in accord with current theories. McDougall (1927, 1930; Rhine and McDougall, 1933) has studied the ability of rats to learn to avoid one of two possible ways of escape from a flooded tank over a period of thirty-four successive generations. The learning curves of the offspring of rats that had had experience of this maze were compared with those of unexperienced controls and a marked improvement was found in the experimental strain, which was obvious from the twelfth generation on. In order to rule out the possibility of involuntary selection in picking out the individuals to be used for breeding, McDougall selected the individuals with the worst learning records for breeding. Improvement of learning capacity was found in the offspring in spite of this counter-selection. This seemed to be the first successful attempt to prove the possibility of Lamarckian evolution in behaviour study. Naturally this work met with considerable criticism.

Crew (1936) took the trouble to repeat McDougall's experiments. He kept in touch with McDougall during the work, used the same (or even improved) methods, and the same strain of rats. Yet he got quite different results and could not find evidence of improvement. The cause of this discrepancy in facts has remained obscure, and in the opinion of Crew the question still remains open. Recently, however, Agar (1950) has shown McDougall's conclusion to be invalid.

THE PART PLAYED BY BEHAVIOUR IN THE CAUSATION OF EVOLUTION

Although this chapter is devoted to the problem of behaviour as a result of evolution, I will insert a few words about some effects of behaviour that, while apparently not important in ecological respect (and for that reason not treated extensively in Chapter VII), deserve attention because of the important part they play in the causation of evolutionary change.

Isolating Mechanisms

In the chapter on the adaptiveness of behaviour I mentioned briefly the fact that 'survival' has a double meaning. It may mean survival in the sense of continuation of life in general, but it may also have the more restricted meaning of survival as a distinct species. Although most of the discussions in Chapter VII centred around the wider concept, survival value in the narrower sense interests us here because survival as a distinct species is an important aspect of speciation. As mentioned on p. 198, certain aspects of behaviour play the part of isolating mechanisms preventing the interbreeding of species.

The studies undertaken thus far indicate that isolating mechanisms sometimes depend on processes taking place during or after coition, sometimes on processes occurring earlier. Behaviour characteristics play a particularly important part in the latter. As a matter of fact, sexual isolation is one of the important functions of courtship behaviour, the other functions, as discussed in Chapter VII, being the synchronization, releasing, and directing of sexual responses in the mate.

The nature of sexual isolation between *Drosophila pseudo-obscura* and *D. persimilis* has been studied by Mayr (1946). Although males of *D. pseudo-obscura* when deprived of their antennae (the organs of smell) were distinctly less discriminating than were intact males, the main obstacle to interbreeding is the fact that interspecific copulations are usually incomplete, the male withdrawing its phallus after an insertion of only 1–2 seconds.

Spieth (1947) studied the courtship of six of the seven closely related species of the *Drosophila willistoni* group. The males approach individuals of either sex and of either species of the group. Before mounting the female, the male taps her side with his fore legs. If the female belongs to another species, all the males of every species of this group break off courtship at this point. This indicates that the stimulus received by the male at this point is highly specific and that the selectivity of the IRM of the next link in the male's reaction acts as a sexual isolating mechanism. In other words, we have to do with a typical releaser-relationship. It is very probable that sexual isolation is often brought about in a similar way.

Such a state of affairs as Spieth's work brings to light in the *Drosophila willistoni* group may be exceptional in that it is unusually simple. Considering the fact that innate responses are released by relatively simple sign stimuli and go off sometimes in the wrong situation, one might wonder whether one single sign stimulus is not usually too unspecific to ensure sexual isolation. As a matter of fact, the selectivity of mating behaviour is not, as a rule, dependent on only one IRM but on a whole series. The ejaculation of sperm in the male *Gasterosteus aculeatus* is dependent on the successive administration of (1) a visual stimulus situation (pregnant female moving in a special way), (2) another visual situation (female following), (3) a third visual situation (female in nest), and (4) a (probably chemical) stimulus provided by the eggs. While each of the IRMs responding to these sign stimuli might be unselective, together they render it highly improbable that a female of another species will ever release the final ejaculation of sperm. In the same way, the chain character of the female's mating behaviour ensures its spawning into nests of its own species.

These considerations might help us to understand a peculiarity of a type of courtship behaviour found in some insects and possibly in other animals as well. In the grayling butterfly the female, after alighting in response to the male's aerial pursuit, keeps motionless during the lengthy courtship performance of the male. Analysis of this courtship shows that it consists of a chain of different reactions, each of them responding to different stimuli from the female. Yet the female does not display; it just sits, and throughout the male's display continuously presents all the stimuli to which the male successively responds. It is as if the male, being unable by the very nature of innate responsiveness to respond to the great number of stimuli necessary to render him sufficiently selective, must reach this high level of selectivity by reacting to a series of different sign stimuli.

It would be highly interesting to study various types of courtship from this point of view. It might well be that several characteristics that cannot be understood merely as either synchronization, releasing, or directing devices may act as isolating mechanisms.

Selection

Behaviour plays a part in selection in two ways: intraspecifically and interspecifically. Intraspecifically, mating, fighting, care of the young, reactions of the young to parents may all be selective towards individual members of the species. Interspecifically, predators may select their prey, and in general, competition affects behaviour as well as other adaptive characters.

Reliable evidence on intraspecific selection is scarce. Anecdotal observations about preferential mating are usually not entirely conclusive. Selous's well-known observations (1906–7) on preferential mating in the ruff might be mentioned as an example. Selous made daily observations on the mating behaviour of a group of ruffs at one particular 'lek' over a long period. He found that certain males (and especially those with highly developed epigamic characters) were chosen by females much more often than others and he considered this to be a proof of sexual selection. Although his observations might well be interpreted in this way, it is not entirely impossible (although admittedly not very probable) that the females reacted to the particular mating-station rather than to the male occupying it. The males tend to occupy the same station day after day.

A more convincing study is that made by Noble and Curtis (1939) on the jewel fish, *Hemichromis maculatus*. These authors carried out a number of selection tests of the following type. A female in sexually active condition was placed in a tank between two other tanks, in each of which a male was placed. In one of the males the sexual colour pattern was intensified by treatment with yohimbine. The females showed preference for the more brilliantly coloured males by settling near them.

Ethologically this result is not surprising. The existence of sexual selection has been doubted and even denied by animal psychologists on the ground that the psychological level of many animals does not allow them to make a deliberate choice. However, the importance of sexual selection as an agent in evolution lies in the result, and the result of preferential reaction to one of two animals or situations can be effected without 'deliberate' or 'conscious' choice. On the level of innate reactivity to sign stimuli a quantitative difference in stimulating power between two situations can cause an animal to 'select' one of them merely because this one releases more responses than the other, or releases more complete responses. Even if an animal does not 'choose' in the anthropomorphic sense of the word, the result may nevertheless be 'selection'. As slight differences in stimulating value of males and females are always found, there is not the slightest doubt that intraspecific selection is of wide occurrence. Yet more experimental studies are badly needed, because it is necessary to know how detailed the selection is, how finely it works.

It has often been stressed (Haldane, 1932; Huxley, 1940) that intraspecific selection may lead to the excessive development of organs (hypertely), which, though of advantage in one respect, may be distinctly disadvantageous in another. Ethology may help us to understand the exact nature of this peculiar process, which may have led to the extinction of a number of species. It is a striking fact that some of the classical examples of hypertely, such as the enormous antlers of *Megaceros hibernicus*, the giant deer, and the canines of *Smilodon*, the sabre-toothed tiger, are organs of a type that would be supposed to be social releasers by the present-day ethologist. We have seen in Chapter VII that releasers, though they may serve various ends, have in common that they all have selective value as means of facilitating social co-operation. Our present knowledge of social releasers justifies the belief that hypertely may, much more often than is generally believed, concern social releasers rather than functionless structures or 'luxuries'.

Behaviour as an agent of interspecific selection is of extreme importance in the relationship of predator and prey. As one example I choose visual adaptations, and especially cryptic coloration, the high adaptive value of which has long been affirmed by some biologists and resolutely denied by others. Experiments on the survival value of cryptic coloration with its accessories, shape, and behaviour (see p. 165), are important in various respects, two of which concern us here. First, it has only recently been proved by reliable experiments that cryptic coloration has any survival value at all (Steiniger, 1938; Sumner, 1934, 1935*a*, 1935*b*; Isely, 1938; Dice, 1947, and others). However, most experiments carried out thus far have compared the relative vulnerability of a given species in two extremely different situations, viz. one in which the prey's colour matched the environment very well and another in which its colour was in striking contrast to the environment. It is, however, of great interest to know whether much subtler differences also have survival value, or in other words, whether even the slightest deviation from the maximum cryptic effect is perceived by the predator. For it is only if this can be proved experimentally that selection would be judged sufficient to account for the amazing refinement of cryptic adaptations actually found in nature. We possess only indirect evidence pointing in this direction: De Ruiter's observations on European jays show that this species far surpasses man in detecting cryptically coloured prey. Yet animals showing the most extreme cryptic adaptation, such as twig-like larvae (*Ennomos* species) and certain sphingid caterpillars famous for their marvellous countershading (*Smerinthus ocellatus* and others) were too much even for this species, and were protected to a high degree when presented in natural situations (De Ruiter, personal communication).

THE ETHOLOGICAL STUDY OF MAN

Man is an animal. He is a remarkable and in many respects unique species, but he is an animal nevertheless. In structure and functions, of the heart, blood, intestine, kidneys, and so on, man closely resembles other animals, especially other vertebrates. Palaeontology as well as comparative anatomy and embryology do not leave the least doubt that this resemblance is based on true evolutionary relationships. Man and the present-day primates have only recently diverged from a common primate stock. This is why comparative anatomy and comparative physiology have yielded such important results for human biology. It is only natural, therefore, that the zoologist should be inclined to extend his ethological studies beyond the animals to man himself. However, the ethological study of man has not yet advanced very far. While animal neurophysiology and animal ethology are coming in touch with each other, there remains a wide gap between these two fields in the study of behaviour of man.

One of the main reasons for this is the almost universal misconception that the causes of man's behaviour are qualitatively different from the causes of animal behaviour. Somehow it is assumed that only the lowest building-stones of behaviour, such as impulse flow in peripheral nerves, or simple reflexes, can be studied with neurophysiological or, in general, objective methods, while behaviour as an integrated expression of man as a whole is the subject-matter of psychology. Somehow it is assumed that, when, in investigating behaviour, one climbs higher and higher in the hierarchical structure, ascending from reflexes or automatisms to locomotion, from here to the higher level of consummatory acts, and to still higher levels, one will meet a kind of barrier bearing the sign 'Not open to objective study; for psychologists only'. It is of fundamental importance to recognize the utter fallacy of such a conception. As long as neurophysiology focused its attention on lower levels, there was such a gap between the spheres of interest of physiology and of psychology that the existence of the barrier somewhere in no-man's-land could neither be proved nor disproved. But neurophysiology has been including higher and higher levels in its area of work, psychology is beginning to look at the lower, instinctive levels, and ethology has settled in between and so meets neurophysiologists at its lower levels and psychologists at its highest levels. And one of the first results of this expansion is that we now realize that a barrier of this kind does not exist. Neurophysiology and ethology are of the same way of thinking, and they will co-operate and even fuse into one science. Psychology in the most restricted sense, or the study of subjective phenomena, will have an independent existence alongside ethology. The higher levels of nervous function can be studied by both sciences without creating conflict; each one of them will unveil one aspect of the phenomenon. The truth of this conclusion can already be proved on the level of instinctive activity. Food-seeking behaviour, for instance, has been studied by the physiologists. They find that it is a highly integrated muscle action, caused by a system of nervous motor impulses originating in a higher centre in the hypothalamus (see p. 108).

This neural system can be influenced by external stimuli and by internal stimuli dependent on contractions of muscles in the stomach wall (see p. 66). The psychologist is not primarily interested in this objective picture of food-seeking behaviour; he will point to the subjective phenomenon of hunger, and some psychologists even assume that emotions like hunger somehow take part in the objective causation of behaviour. In this assumption, there is a real conflict between psychology and ethology. Psychology does not come into conflict with objective study of the lowest levels such as the reflex level, because introspection does not reach them. At the higher level, introspection brings us into contact with an aspect of behaviour that is out of reach of objective study. We know that both aspects belong to one reality, but somehow the scientist's mind is unable to synthesize them into one harmonious picture. As scientists, we have to recognize the duality of our thinking and to accept it. If we do so, we can proceed with an objective study of human behaviour without running the risk of questioning, with an inexcusable narrowness of mind, the value of the other category of thinking. In other words, both the data gained by introspection and those found by objective study are facts. When the food-seeking drive is aroused, the subject experiences hunger; meanwhile his neural mechanisms are at work. Discovering those neural mechanisms is not discovering that subjective phenomena do not occur.

The objective study of human behaviour meets with one serious difficulty: terminology. The ethologist may be consistent, and create new objective terms for processes already known and named from the subjective point of view. This will often seem a ridiculous procedure. As an illustration, I may quote a colleague of mine who, after a discussion on the physiological aspects of the food-seeking drive, remarked: 'Well, anyhow, I think it is more convenient to say "I am hungry" than to say that the enteroceptors in the muscles of my stomach wall cause impulses to run to my hypothalamus, &c.' The other possibility is to use terms borrowed from everyday language and to say 'I am hungry' as a convenient description, not only of the subjective phenomenon experienced, but also of the physiological state. In doing this, however, the ethologist will always be too late; the psychologist has already claimed such terms and loaded them with subjective meaning, and even when the ethologist uses such terms, explicitly stating that they serve as mere descriptive terms, the psychologist will accuse him of trespassing and will even take it as a proof of the impossibility of objective study of behaviour (see, e.g. Bierens de Haan, 1947). Therefore the greatest caution has to be applied in terminology, as long as the misunderstanding about the aim and scope of the two sciences prevails.

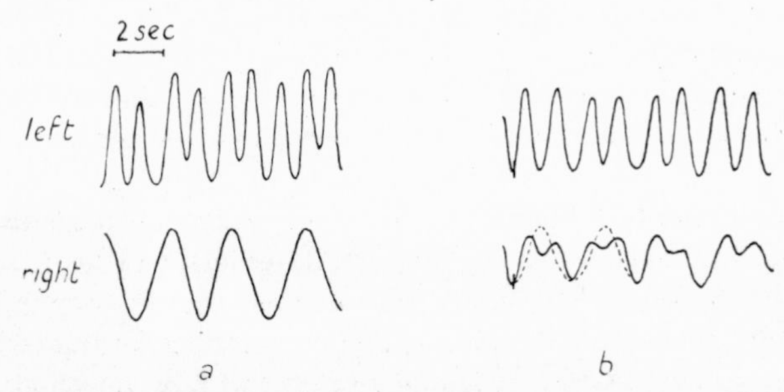

FIG. 129. Relative co-ordination (superposition) in man. (*a*) Above: dependent arm; below: dominant arm. (*b*) Above: dominant arm; below: dependent arm. Further explanation in the text. After von Holst, 1938.

As I said before, human ethology is still in its infancy. Yet the evidence is already much too voluminous to be covered here. All I can do is to pick a few examples of either human or animal behaviour and to show how some processes usually considered as typical of animals are also found in man.

The Lower Levels

That the nervous impulse is a process which is very similar or identical in man and in animals is considered so self-evident that in text-books of human physiology nearly all the concrete facts and illustrations concern nervous excitation in frogs, cats, 'the rabbit', and the like.

The simpler co-ordinations, like reflexes, are also thought to be essentially identical, and even locomotion is not treated differently.

Recently von Holst has analysed some instances of co-ordination between separate rhythms. He shows that one of the principles of relative co-ordination, the principle of superposition, is also found in man, and that man is not able to counteract it through his higher centres. When both arms are swung rhythmically up and down, one arm twice as fast as the other, and the subject tries to keep the amplitude of the faster arm constant, it is found that the faster arm makes long and short sweeps alternately, dependent on whether its movements coincide with a down stroke or with an up stroke of the slower arm (Fig. 129).

Locomotion as a whole has been proved in several animals to be governed by an entirely innate co-ordination pattern (Weiss, 1941*b*, *c*). The improvement observed during growth, when much 'practising' occurs, is not due to this practising but to growth (Carmichael, 1926, 1927; Grohmann, 1939). According to the work of Gesell (1947), McGraw (1947), the same holds good for locomotion in man. The walking pattern is innate and the greater part of the improvement shown in the first years of life is due to growth of the nervous connexions.

Instincts

Instinctive behaviour in man has been studied in its various aspects: motor pattern, internal factors (motivation), and external factors (sensory stimuli).

The motivation of human behaviour is a matter of much discussion. Here again, introspection is a hindrance to understanding: it reveals to us only the conscious subjective phenomena, though we have learnt from Freud that non-conscious phenomena of a quite different nature are at work as well. As a very simple example of how conscious reasoning may entirely distort our insight into the real causes underlying our behaviour, one of the commonest conflicts between drive and reasoning may be taken. Mating behaviour in man, not in the form of the accomplishment of the consummatory act, but in the preparatory, appetitive stage of 'love-making', proves, when studied ethologically, to be basically dependent on sex hormones and on external stimuli and it is on these agents that our rational powers exact a regulating influence. Now every individual among us who has the habit of self-observation, and who has not forgotten his youth, knows how often the urge has driven him 'blindly', 'against his better judgement' to obey it, when there was a conflict between better judgement and drive. Falling in love changes one's entire outlook upon one's surroundings. Criminologists teach us that the number of crimes, even of serious ones like murder, committed in obedience to the instinctive urge to show off before a female is astonishingly high. Quite different, but from our standpoint equally significant, are murders due to sexual rivalry.

In a similar way, the instinctive food-seeking drive often conflicts with reason. This is a rare phenomenon in normal Western society. But everyone who has lived through periods of real starvation—a condition, common enough outside the Western world, that has touched western Europe just long enough to make its significance clear to us—knows how relatively weak reason is when it is up against really powerful instinctive motivation.

The receptive side of instinct has not yet been thoroughly studied in man. Lorenz has shown in an important contribution (1943) that Innate

Releasing Mechanisms are to be found in man as well. Two of them may be mentioned here. The parental instinct, a sub-instinct of the major reproductive instinct, is responsive to sign stimuli provided by the human baby. Although no well-planned experiments with models have been carried out, there is evidence from three sources which has about the same value as such experiments. First, dolls are adapted to meet the

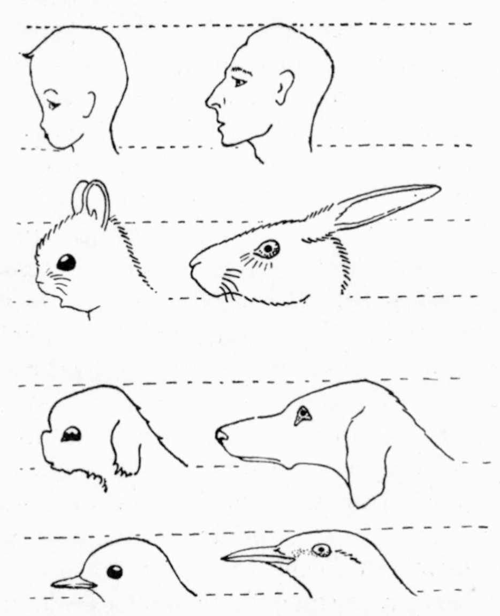

FIG. 130. Human baby and three substitutes (left) presenting sign stimuli that release parental reactions in man, as compared with adult man and three animals (right) which do not release parental conduct. After Lorenz, 1943.

demands of the IRM. Second, the film industry, intending to meet man on the instinctive level, has developed an optimal baby. Third, childless women select substitutes for babies in their pets. A study of substitutes of these three kinds reveals that the human parental instinct responds to the following sign stimuli: a short face in relation to a large forehead, protruding cheeks, maladjusted limb movements. Fig. 130 illustrates the morphological releasers, present in the left-hand figures, absent from the right-hand figures.

Lorenz's second example concerns a non-social response. It further differs from the first in that it concerns a relatively simple stereotyped motor response. The touch stimuli given by an insect crawling on the skin release the response of throwing it off with a quick movement of the hand. This movement contains both a fixed pattern and an orientation component, and is, therefore, a reaction of greater complexity than a mere reflex. It is probably innate and matures relatively late. It is accompanied by a subjective phenomenon, disgust.

Another phenomenon suggesting an instinctive organization in man basically similar to that found in animals is displacement activity. Displacement activities are by no means rare in man. They are not so easily recognized as in animals because in man learned patterns, like lighting a cigarette, handling keys or handkerchief, &c., often act as displacement activities. However, innate patterns may function as outlets in man too. The general occurrence of scratching behind one's ear in conflict situations almost certainly has an innate basis. It is striking how often activities belonging to the instinct of comfort (care of the skin) are shown in conflict situations: in women it mostly takes the form of adjusting non-existing disorder of the coiffure, in the man it consists of handling the beard or moustache, not only in the days when men still had them but also in this 'clean-shaven' era. Further, it is striking that displacement scratching can be observed regularly in primates (Tinbergen, 1939*b*).

Another innate displacement activity in man seems to be sleep. In low intensities, in the form of yawning, it is of common occurrence in mild conflict situations. Just as in some birds (avocet, oystercatcher, and other waders) actual sleep is an outlet in situations where the aggressive instinct and the instinct of escape are simultaneously aroused. Reliable and trained observers, among them Professor P. Palmgren of Helsingfors, have told me that in situations of extreme tension at the front, just before an actual attack, infantrymen may be overcome by a nearly insurmountable inclination to go to sleep. Sleep, as is known from Hess's experiments, is a true instinctive act, depending on stimulation of a centre in the hypothalamus. It is also in line with other instinctive acts in that it is the goal of a special kind of appetitive behaviour.

Still another innate displacement activity in man seems to be the occurrence of sperm-ejaculation as a consequence of a blocked escape drive at examinations (Bilz, 1941).

Because of the variability of displacement activities in man, which is the consequence of the variability of his instinctive behaviour as a whole, displacement activities are difficult to study in man. Yet here is a problem, the study of which could contribute much to a better understanding of the part played by instincts in the behaviour of man.

BIBLIOGRAPHY

ABEL, O., 1912: *Grundzüge der Palaeobiologie der Wirbeltiere*. Stuttgart.

ADRIAN, E. D., 1931: Potential changes in the isolated nervous system of *Dytiscus marginalis*. *J. Physiol.* **72**, 132–51.

—— 1932: *The Mechanism of Nervous Action*. Philadelphia.

—— and F. J. J. BUYTENDIJK, 1931: Potential changes in the isolated brain of the goldfish. *J. Physiol.* **71**, 121–35.

ADRIAANSE, A., 1947: *Ammophila campestris* Latr. und *Ammophila adriaansei* Wilcke. *Behaviour*, **1**, 1–35.

AGAR, W. E., F. H. DRUMMOND, and D. W. TIEGS, 1948: Third report on a test of McDougall's Lamarckian experiment on the training of rats. *J. Exp. Biol.* **25**, 103–22.

ALLEN, A. A., 1934: Sex rhythm in the Ruffed Grouse (*Bonasa umbellus* L.) and other birds. *Auk*, **51**, 180–99.

ARMSTRONG, E. A., 1947: *Bird Display and Behaviour*. Cambridge.

AUTRUM, H., 1940: Über Lautäußerung und Schallwahrnehmung bei Arthropoden. II. *Zs. vergl. Physiol.* **28**, 326–52.

—— and H. STUMPF, 1950: Das Bienenauge als Analysator für polarisiertes Licht. *Zs. f. Naturf.* 5b, 116–122.

BAERENDS, G. P., 1939: Waarnemingen en proeven aan de ruggezwemmer (*Notonecta glauca*). *De Levende Natuur*, **44**, 11–17, 45–51.

—— 1941: Fortpflanzungsverhalten und Orientierung der Grabwespe *Ammophila campestris* Jur. *Tijdschr. Entomol.* **84**, 68–275.

—— and J. M. BAERENDS, 1950: An introduction to the study of the ethology of Cichlid Fishes. *Behaviour*, Supplement, **1**, 1–242.

BALDUS, K., 1926: Experimentelle Untersuchungen über die Entfernungslokalisation der Libellen (*Aeschna cyanea*). *Zs. vergl. Physiol.* **3**, 475–506.

BALL, J., 1934: Sex behavior of the rat after removal of the uterus and vagina. *J. Comp. Psychol.* **18**, 419–22.

BEACH, F. A., 1942: Analysis of factors involved in the arousal, maintenance and manifestation of sexual excitement in male animals. *Psychosom. Med.* **4**, 173–98.

—— 1943: Importance of progesterone to induction of sexual receptivity in spayed female rats. *Proc. Soc. Exp. Biol. Med.* **51**, 369–71.

—— 1948: *Hormones and Behaviour*. New York.

DE BEER, G. R., 1940: *Embryos and Ancestors*. Oxford.

BEUSEKOM, G. VAN, 1946: *Over de orientatie van de Bijenwolf* (*Philanthus triangulum* Fabr.). Leiden.

—— 1948: Some experiments on the optical orientation in *Philanthus triangulum* Fabr. *Behaviour*, **1**, 195–225.

BIERENS DE HAAN, J. A., 1937: Über den Begriff des Instinktes in der Tierpsychologie. *Folia Biotheoretica*, **2**, 1–16.

—— 1947: Animal psychology and the science of animal behaviour. *Behaviour*, **1**, 71–80.

BILZ, R., 1941: Zur Psychophysik des Verlegenheitskratzens. *Zentralbl. Psychother. und ihre Grenzgeb.* **13**, 36–50.

BOESEMAN, M. J., J. VAN DER DRIFT, J. M. VAN ROON, N. TINBERGEN, and J. J. TER PELKWIJK, 1938: De bittervoorns en hun mossels. *De Levende Natuur*, **43**, 129–36.

BOLS, J., 1935: Opruimingswerk der watertorlarf *Hydrophilus piceus* L. *De Levende Natuur*, **40**, Gedenkboek Dr. J. P. Thysse.

BOSS, W. R., 1943: Hormonal determination of adult characters and sex behavior in Herring Gulls (*Larus argentatus*). *J. Exp. Zool.* **94**, 181–220.

BROCK, FR., 1926: Das Verhalten des Einsiedlerkrebses *Pagurus arrosor* Herbst während der Suche und Aufnahme der Nahrung. *Zs. Morphol. Ökol. Tiere*, **6**, 415–552.

—— 1936: Suche, Aufnahme und enzymatische Spaltung der Nahrung durch die Wellhornschnecke *Buccinum undatum* L. *Zoologica* (Wien), **34**, 1–136.

BRÜCKNER, G. H., 1933: Untersuchungen zur Tiersoziologie, insbesondre der Auflösung der Familie. *Zs. Psychol.* **128**, 1–120.

BRÜGGER, M., 1943: Freßtrieb als hypothalamisches Symptom. *Helv. Physiol. Acta*, **1**, 183–98.

BRUN, R., 1914: *Die Raumorientierung der Ameisen und das Orientierungsproblem im Allgemeinen*. Jena.

BUDDENBROCK, W. VON, 1931: Untersuchungen über den Schattenreflex. *Zs. vergl. Physiol.* **13**, 164–214.

CARMICHAEL, L., 1926: The development of behavior in vertebrates experimentally removed from the influence of external stimulation. *Psychol. Rev.* **33**, 51–8.

—— 1927: A further study of the development of behavior in vertebrates experimentally removed from the influence of external stimulation. Ibid. **34**, 34–47.

CARPENTER, C. R., 1933: Psychobiological studies of social behavior in Aves. II. The effect of complete and incomplete gonadectomy on secondary sexual activity, with histological studies. *J. Comp. Psychol.* **16**, 59–98.

CATE, J. TEN, 1939: Zur Frage der rhythmischen Tätigkeit des Rückenmarks bei Haifischen. *Arch. Néerl. Physiol.* **24**, 226–41.

CHASE, P. E., 1940: An experimental study of the relations of sensory control to motor function in amphibian limbs. *J. Exp. Zool.* **83**, 61–87.

CINAT-TOMSON, H., 1926: Die geschlechtliche Zuchtwahl beim Wellensittich (*Melopsittacus undulatus* Shaw). *Biol. Zbl.* **46**, 543–52.

COGHILL, G. E., 1929: *Anatomy and the Problem of Behaviour*. Cambridge.

CRAIG, W., 1911: Oviposition induced by the male in pigeons. *J. Morphol.* **22**, 299–305.

—— 1918: Appetites and aversions as constituents of instincts. *Biol. Bull.* **34**, 91–107.

CRANE, J., 1941: Crabs of the genus *Uca* from the West coast of Central America. *Zoologica N.Y.* **26**, 145–208.

CREED, R. S., D. DENNY-BROWN, J. C. ECCLES, E. G. T. LIDDELL, and C. S. SHERRINGTON, 1932. *Reflex Activity of the Spinal Cord.* Oxford.

CREW, F. A. E., 1936: A repetition of McDougall's Lamarckian experiment. *J. Gen.* **33**, 61–79.

DAVIS, D. E., 1942: The phylogeny of social nesting habits in the Crotophaginae. *Quart. Rev. Biol.* **17**, 115–34.

—— and L. V. DOMM, 1943: The influence of hormones on the sexual behavior of domestic fowl. *Essays in Biology, in Honor of Herbert M. Evans*, 171–81. Univ. of California Press.

DELACOUR, J., and E. MAYR, 1945: The family Anatidae. *Wilson Bull.* **57**, 3–55.

DEMPSEY, E. W., 1939: The relationship between the central nervous system and the reproductive cycle in the female guinea pig. *Amer. J. Physiol.* **126**, 758–65.

DETWILER, S. R., and R. H. VAN DYKE, 1934: The development and function of deafferented forelimbs in Amblystoma. *J. Exp. Zool.* **68**, 321–46.

DEWAR, J. M., 1908: Notes on the oystercatcher (*Haematopus ostralegus*), with reference to its habit of feeding upon the mussel (*Mytilus edulis*). *Zoologist* (4), **12**, 201–12.

—— 1913: Further observations on the feeding habits of the oystercatcher (*Haematopus ostralegus*). Ibid. (4), **17**, 41–56.

DICE, L. R., 1945: Minimum intensities of illumination under which owls can find dead prey by sight. *Amer. Natural.* **79**, 385–416.

—— 1947: Effectiveness of selection by owls of deer-mice (*Peromyscus maniculatus*) which contrast in color with their background. *Contrib. Lab. Vertebr. Biol. Ann Arbor.* **34**, 1–20.

DIJKGRAAF, Sv., 1934: Untersuchungen über die Funktion der Seitenorgane an Fischen. *Zs. vergl. Physiol.* **18**, 65–112.

—— 1946: Die Sinneswelt der Fledermäuse. *Experientia*, **2**, 438–49.

DUYM, M. and G. M. VAN OYEN, 1948: Het sjirpen van de zadelsprinkhaan. *De Levende Natuur*, **51**, 81–7.

EDWARDS, G., E. HOSKING, and STUART SMITH, 1948: Aggressive display of the Oystercatcher. *Brit. Birds*, **41**, 236–43.

EMLEN, JOHN T., and F. W. LORENZ, 1942: Pairing response of free-living valley quail to sex hormone pellet implants. *Auk*, **59**, 369–78.

ENGELMANN, W., 1928: Untersuchungen über die Schallokalisation bei Tieren. *Zs. Psychol.* **105**, 317–70.

EVANS, L. T., 1935: Winter mating and fighting behavior of *Anolis carolinensis* as induced by pituitary injections. *Copeia*, 1935, 3–6.

FABRE, J. H., 1923: *Souvenirs entomologiques*. 80th ed. Paris.

FRAENKEL, G., 1931: Die Mechanik der Orientierung der Tiere im Raum. *Biol. Rev.* **6**, 36–87.

—— and D. L. GUNN, 1940: *The Orientation of Animals*. Oxford.

FRASER DARLING, F., 1938: *Bird Flocks and the Breeding Cycle*. Cambridge.

FREDERIKS, H. H. J., 1941: Proeven over hormonale beïnvloeding van instinctieve geslachtsuitingen van vrouwelijke kanaries na inspuiting van testosteron propionaat. *Tijdschr. Diergeneesk.* **68**, 537–8.

FRIEDLÄNDER, B., 1894: Beiträge zur Physiologie des Zentralnerven-systems und des Bewegungsmechanismus der Regenwürmer. *Pflüger's Archiv.* **58**, 168–206.

FRIEDMANN, H., 1929: *The Cowbirds*. Springfield-Baltimore.

FRISCH, K. VON, 1914: Der Farbensinn und Formensinn der Biene. *Zool. Jahrb. Allg. Zool. Physiol.* **35**, 1–188.

—— 1923: Über die 'Sprache' der Bienen. *Zool. Jahrb. Allg. Zool. Physiol.* **40**, 1–186.

—— 1926: Vergleichende Physiologie des Geruchs- und Geschmackssinnes. *Handb. norm. pathol. Physiol.* **11**, 1, 203–40.

—— 1927: *Aus dem Leben der Bienen*. Berlin.

—— 1934: Über den Geschmackssinn der Biene. *Zs. vergl. Physiol.* **21**, 1–156.

—— 1946: Die Tänze der Bienen. *Österr. Zool. Zs.* **1**, 1–48.

—— 1950: Die Sonne als Kompass im Leben der Bienen. *Experientia*, **6**, 210–21.

FRISCH, K. VON and H. STETTER, 1932: Untersuchungen über den Sitz des Gehörsinnes bei der Elritze. *Zs. vergl. Physiol.* **17,** 686–802.

GENTZ, K., 1935: Zur Brutpflege des Wespenbussards. *J. Ornithol.* **83,** 105–15.

GESELL, A., 1947: The ontogenesis of infant behavior. In: *Manual of Child Psychology*, New York–London.

GOETHE, F., 1937: Beobachtungen und Erfahrungen bei der Aufzucht von deutschem Auerwild. *Deutsche Jagd.*

GRAHAM BROWN, 1911: The intrinsic factors in the act of progression in the Mammal. *Proc. Roy. Soc. London*, B, **84,** 308–20.

—— 1912: The factors in rhythmic activity of the nervous system. Ibid. B, **85,** 278–89.

GRAY, J., 1936: Studies on animal locomotion. IV. The neuromuscular mechanism of swimming in the eel. *J. Exp. Biol.* **13,** 170–80.

—— 1939: Aspects of animal locomotion. *Proc. Roy. Soc. London*, B, **128,** 28–62.

—— and H. W. LISSMANN, 1940: The effect of deafferentation upon the locomotory activity of amphibian limbs. *J. Exp. Biol.* **17,** 227–36.

GRETHER, W. F., 1939: Color vision and color blindness in monkeys. *Comp. Psychol. Monogr.* **15,** 1–38.

GROHMANN, J., 1939: Modifikation oder Funktionsreifung? *Zs. Tierpsychol.* **2,** 132–44.

HALDANE, J. B. S., 1932: *The Causes of Evolution*. London.

HECHT, S., 1934: The nature of the photoreceptor process. *Handb. Gen. Exp. Psychol.*

—— and M. H. PIRENNE, 1940: The sensibility of the long-eared owl in the spectrum. *J. Gen. Physiol.* **23,** 709–17.

—— and E. WOLF, 1929: The visual acuity of the honey bee. Ibid. **12,** 727–61.

HEILBORN, A., 1930: *Liebesspiele der Tiere*. Berlin–Charlottenburg.

HEINROTH, O., 1909: Beobachtungen bei der Zucht des Ziegenmelkers (*Caprimulgus europaeus* L.). *J. Ornithol.* **57,** 56–83.

—— 1911: Beiträge zur Biologie, namentlich Ethologie und Psychologie der Anatiden. *Verh. V. Int. Ornithol. Kongr.* Berlin.

—— and M. HEINROTH, 1928: *Die Vögel Mitteleuropas*. Berlin.

HEMMINGSEN, A. M., 1933: Studies on the oestrus-producing hormone. *Skand. Arch. Physiol.* **65,** 97–250.

HERTER, K., 1933: Dressurversuche mit Igeln. *Zs. vergl. Physiol.* **18,** 481–516.

—— 1941: Die Vorzugstemperaturen bei Landtieren. *Naturwiss.* **29,** 155–64.

HERTZ, M., 1929: Die Organisation des optischen Feldes bei der Biene. I. *Zs. vergl. Physiol.* **8,** 693–748.

—— 1930: Die Organisation des optischen Feldes bei der Biene. II. *Zs. vergl. Physiol.* **11,** 107–45.

—— 1931: Die Organisation des optischen Feldes bei der Biene. III. Ibid. **14,** 629–74.

HESS, C. VON, 1913: Experimentelle Untersuchungen über den angeblichen Farbensinn der Bienen. *Zool. Jahrb. Allg. Zool. Physiol.* **34,** 81–106.

HESS, W. R., 1943: Das Zwischenhirn als Koordinationsorgan. *Helv. Physiol. Acta*, **1,** 549–65.

—— 1944: Das Schlafsyndrom als Folge dienzephaler Reizung. Ibid. **2,** 305–44.

—— and M. BRÜGGER, 1943: Das subkortikale Zentrum der affektiven Abwehrreaktion. Ibid. **1,** 33–52.

HESS, W. R. and M. BRÜGGER, 1943: Der Miktions- und der Defäkationsakt als Erfolg zentraler Reizung. Ibid. **1**, 511–33.

—— and W. O. C. MAGNUS, 1943: Leck- und Kau-Automatismen bei elektrischer Reizung im Zwischenhirn. Ibid. **1**, 533–47.

HOFFMANN, H., 1930: Über den Fluchtreflex bei *Nassa*. *Zs. vergl. Physiol.* **11**, 662–88.

HOLST, E. VON, 1934: Studien über die Reflexe und Rhythmen beim Goldfisch (*Carassius auratus*). *Zs. vergl. Physiol.* **20**, 582–99.

—— 1935*a*: Über den Lichtrückenreflex bei Fischen. *Pubbl. Staz. Zool. Napoli*, **25**, 143–58.

—— 1935*b*: Erregungsbildung und Erregungsleitung im Fischrückenmark. *Pflüger's Archiv*, **235**, 345–59.

—— 1935*c*: Über den Prozeß der zentralnervösen Koordination. Ibid. **236**, 149–58.

—— 1936: Versuche zur Theorie der relativen Koordination. Ibid. **237**, 93–121.

—— 1937: Vom Wesen der Ordnung im Zentralnervensystem. *Naturwiss.* **25**, 625–31, 641–7.

—— 1938: Über relative Koordination bei Säugern und beim Menschen. *Pflüger's Arch.* **240**, 44–59.

—— 1941: Entwurf eines Systems der lokomotorischen Periodenbildungen bei Fischen. Ein kritischer Beitrag zum Gestaltproblem. *Zs. vergl. Physiol.* **26**, 481–529.

HOLZAPFEL, M., 1940: Triebbedingte Ruhezustände als Ziel von Appetenzverhalten. *Naturwiss.* **28**, 273–80.

HOWARD, H. E., 1907–15: *The British Warblers*. 9 parts. London.

—— 1920: *Territory in Bird Life*. London.

—— 1929: *An Introduction to the Study of Bird Behaviour*. Cambridge.

HUXLEY, J. S., 1914: The courtship habits of the Great crested Grebe (*Podiceps cristatus*); with an addition to the theory of sexual selection. *Proc. zool. Soc. London*, 491–562.

—— 1934: A natural experiment on the territorial instinct. *Brit. Birds*, **27**, 270–7

—— 1940: *Evolution, the Modern Synthesis*. London.

ILSE, D., 1929: Über den Farbensinn der Tagfalter. *Zs. vergl. Physiol.* **8**, 658–92.

ISELY, F. D., 1938: Survival value of acridian protective coloration. *Ecology*, **19**, 370–89.

JENNINGS, H. S., 1923: *The Behavior of the Lower Organisms*. New York.

KATZ, D. and G. RÉVÉSZ, 1909: Experimentell-psychologische Untersuchungen mit Hühnern. *Zs. Psychol.* **50**, 51–9.

KIRKMAN, F. B., 1937: *Bird Behaviour*. London–Edinburgh.

KITZLER, G., 1941: Die Paarungsbiologie einiger Eidechsen. *Zs. Tierpsychol.* **4**, 353–402.

KLUYVER, H. N., 1947: Over het gedrag van een jonge Grauwe Vliegenvanger en van een troep Pestvogels in de winter. *Ardea*, **35**, 131–5.

KNÖLL, F., 1921–26: *Insekten und Blumen*. Wien.

KOEHLER, O., 1932: Beiträge zur Sinnesphysiologie der Süßwasserplanarien. *Zs. vergl. Physiol.* **16**, 606–756.

—— and A. ZAGARUS, 1937: Beiträge zum Brutverhalten des Halsbandregenpfeifers (*Charadrius h. hiaticula* L.). *Beitr. Fortpfl. biol. Vögel.* **13**, 1–9.

KÖHLER, W., 1918: Nachweis einfacher Strukturfunktionen beim Schimpansen und beim Haushuhn. *Abh. Preuß. Akad. Wiss. Phys.-mathem. Kl.* **2,** 3–101.

KORTLANDT, A., 1940*a*: Wechselwirkung zwischen Instinkten. *Arch. néerl. Zoöl.* **4,** 442–520.

—— 1940*b*: Eine Übersicht der angeborenen Verhaltensweisen des Mitteleuropäischen Kormorans (*Phalacrocorax carbo sinensis* (Shaw and Nodder)), ihre Funktion, ontogenetische Entwicklung und phylogenetische Herkunft. Ibid. **4,** 401–42.

KRAMER, G., 1933: Untersuchungen über die Sinnesleistungen und das Orientierungsverhalten von *Xenopus laevis* Laur. *Zool. Jahrb. Allg. Zool. Physiol.* **52,** 630–76.

—— 1937: Beobachtungen über Paarungsbiologie und soziales Verhalten von Mauereidechsen. *Zs. Morphol. Oekol. Tiere,* **32,** 752–84.

KRÄTZIG, H., 1940: Untersuchungen zur Lebensweise des Moorschneehuhns *Lagopus l. lagopus,* während der Jugendentwicklung. *J. Ornithol.* **88,** 139–66.

KRIJGSMAN, B. J., 1930: Reizphysiologische Untersuchungen an blutsaugenden Arthropoden im Zusammenhang mit ihrer Nahrungswahl. I. *Stomoxys calcitrans. Zs. vergl. Physiol.* **11,** 702–30.

KÜHN, A., 1919: *Die Orientierung der Tiere im Raum.* Jena.

—— 1927: Über den Farbensinn der Bienen. *Zs. vergl. Physiol.* **5,** 762–801.

—— 1939: *Grundriß der allgemeinen Zoologie.* Leipzig.

LACK, D., 1939: The display of the blackcock. *Brit. Birds,* **32,** 290–303.

—— 1940: The releaser concept in bird behaviour. *Nature,* **145,** 107.

—— 1940: Courtship feeding in birds. *Auk,* **57,** 169–79.

—— 1943: *The Life of the Robin.* London.

—— 1945: The ecology of closely related species with special reference to cormorant (*Phalacrocorax carbo*) and shag (*P. aristotelis*). *J. Anim. Ecol.* **14,** 12–16.

—— 1946: Competition for food by birds of prey. Ibid. **15,** 123–9.

LASHLEY, K. S., 1938: Experimental analysis of instinctive behaviour. *Psychol. Rev.* **45,** 445–71.

LAVEN, H., 1940: Beiträge zur Biologie des Sandregenpfeifers (*Charadrius hiaticula* L.). *J. Ornithol.* **88,** 183–288.

LEINER, M., 1929: Oekologische Studien an *Gasterosteus aculeatus* L. *Zs. Morphol. Oekol. Tiere,* **14,** 360–400.

—— 1930: Fortsetzung der oekologischen Studien an *Gasterosteus aculeatus* L. Ibid. **16,** 499–541.

LISSMANN, H. W., 1946: The neurological basis of the locomotory rhythm in the spinal dogfish (*Scyllium canicula, Acanthias vulgaris*). I. Reflex behaviour. *J. Exp. Biol.* **23,** 143–61. II. The effect of deafferentiation. Ibid. 162–76.

LOCHER, CH. J. S., 1933: Untersuchungen über den Farbensinn der Eichhörnchen. *Tijdschr. Ned. Dierk. Veren.* **3,** 167–219.

LOEB, J., 1918: *Forced Movements, Tropisms and Animal Conduct.* Philadelphia–London.

LORENZ, K., 1927: Beobachtungen an Dohlen. *J. Ornithol.* **75,** 511–19.

—— 1931: Beiträge zur Ethologie sozialer Corviden. Ibid. **79,** 67–120.

—— 1932: Betrachtungen über das Erkennen der arteigenen Triebhandlungen der Vögel. Ibid. **80,** 50–98.

LORENZ, K., 1935: Der Kumpan in der Umwelt des Vogels. Ibid. **83**, 137–213, and 289–413.

—— 1937*a*: Über den Begriff der Instinkthandlung. *Folia Biotheor.* **2**, 18–50.

—— 1937*b*: Über die Bildung des Instinktbegriffs. *Die Naturwissenschaften*, **25**, 289–300, 307–18, 324–31.

—— 1939: Vergleichende Verhaltensforschung. *Zool. Anz.* Suppl. Bd. **12**, 69–102.

—— 1941: Vergleichende Bewegungsstudien an Anatinen. *J. Ornithol.* **89**, Sonderheft. 19–29.

—— 1943: Die angeborenen Formen möglicher Erfahrung. *Zs. Tierpsychol.* **5**, 235–409.

—— and N. TINBERGEN, 1938: Taxis und Instinkthandlung in der Eirollbewegung der Graugans I. Ibid. **2**, 1–29.

LOTMAR, R., 1933: Neue Untersuchungen über den Farbensinn der Bienen, mit besonderer Berücksichtigung des Ultravioletts. *Zs. vergl. Physiol.* **19**, 673–724.

MCDOUGALL, W., 1927: An experiment for the testing of the hypothesis of Lamarck. *Brit. J. Psychol.* **17**, 267–304.

—— 1930: Second report on a Lamarckian experiment. Ibid. **20**, 201–18.

—— 1933: *An Outline of Psychology*. 6th ed. London.

MCGRAW, M. B., 1947: Maturation of Behavior. In *Manual of Child Psychology*, New York–London.

MAIDL, FR., 1934: *Lebensgewohnheiten und Instinkte der staatenbildenden Insekten*. Wien.

MAIER, N. R. F., and T. L. SCHNEIRLA, 1935: *Principles of Animal Psychology*. New York–London.

MAKKINK, G. F., 1931: Die Kopulation der Brandente (*Tadorna tadorna* (L.)). *Ardea*, **20**, 18–22.

—— 1936: An attempt at an ethogram of the European avocet (*Recurvirostra avosetta* L.) with ethological and psychological remarks. Ibid. **25**, 1–60.

MAST, S. O., 1911: *Light and the Behavior of Organisms*. New York–London.

MATTHES, E., 1932*a*: Geruchsdressuren beim Meerschweinchen. *Zs. vergl. Physiol.* **16**, 766–89.

—— 1932*b*: Weitere Geruchsdressuren beim Meerschweinchen. Ibid. **17**, 464–91.

—— 1948: *Amicta febretta*. Ein Beitrag zur Morphologie und Biologie der Psychiden. *Memór. e estudos do Mus. Zool. Coimbra*, 184, 1–80.

MAYR, E., 1940: Speciation phenomena in birds. *Amer. Natural.* **74**, 249–78.

—— 1942: *Systematics and the Origin of Species*. New York.

—— 1946: Experiments on sexual isolation in *Drosophila*. VII. The nature of the isolating mechanisms between *Drosophila pseudoobscura* and *D. persimilis*. *Proc. Nat. Acad. Sci.* **32**, 128–37.

—— 1948: Ecological factors in speciation. *Evolution*, **1**, 263–88.

MEISENHEIMER, J., 1921: *Geschlecht und Geschlechter im Tierreich*. Jena.

MEYKNECHT, J., 1941: Farbensehen und Helligkeitsunterscheidung beim Steinkauz (*Athene noctua vidalii* A. E. Brehm). *Ardea*, **30**, 129–74.

MOSTLER, G., 1935: Beobachtungen zur Frage der Wespenmimikry. *Zs. Morphol. Oekol. Tiere*, **29**, 381–455.

NALBANDOV, A. W., and L. E. CARD, 1945: Endocrine identification of the broody genotype of cocks. *J. Hered.* **36**, 35–9.

NICE, M. M., 1937: Studies in the life history of the Song Sparrow. I. *Trans. Linn. Soc. N.Y.* **4**.

NICE, M. M., 1943: Studies in the life history of the Song Sparrow. II. Ibid. **6.**

NOBLE, G. K., 1934: Experimenting with the courtship of lizards. *Nat. Hist.* **34,** 1–15.

—— and H. T. BRADLEY, 1933: The mating behavior of lizards. *Ann. N.Y. Acad. Sci.* **35,** 25–100.

—— and B. CURTIS, 1939: The social behavior of the jewel fish, *Hemichromis bimaculatus* Gill. *Bull. Amer. Mus. Nat. Hist.* **76,** 1–46.

—— and B. GREENBERG, 1940: Testosterone propionate, a bisexual hormone in the American Chameleon. *Proc. Soc. Exp. Biol. Med.* **44,** 460–2.

—— and M. WURM, 1940: The effect of testosterone propionate on the black-crowned night heron. *Endocrinology,* **26,** 837–50.

—— and A. ZITRIN, 1942: Induction of mating behavior in male and female chicks following injections of sex hormones. Ibid. **30,** 327–34.

NOLL, H., and E. TOBLER, 1924: *Sumpfvogelleben.* Leipzig.

PAVLOV, J. P., 1926: *Die höchste Nerventätigkeit (das Verhalten) von Tieren.* München.

PEITZMEIER, J., 1942: Die Bedeutung der ökologischen Beharrungstendenz für faunistische Untersuchungen. *J. Ornithol.* **90,** 311–22.

PELKWIJK, J. J. TER, and N. TINBERGEN, 1937: Eine reizbiologische Analyse einiger Verhaltensweisen von *Gasterosteus aculeatus* L. *Zs. Tierpsychol.* **1,** 193–204.

PETERS, H., 1937: Experimentelle Untersuchungen über die Brutpflege von *Haplochromis multicolor*, einem maulbrütenden Knochenfisch. Ibid. **1,** 201–18.

PORTIELJE, A. F. J., 1925: Zur Ethologie bzw. Psychologie der *Rhea americana* L. *Ardea,* **14,** 1–15.

PUKOWSKI, E., 1933: Oekologische Untersuchungen an *Necrophorus* F. *Zs. Morphol. Oekol. Tiere,* **27,** 518–86.

RÄBER, H., 1948: Analyse des Balzverhaltens eines domestizierten Truthahns (*Meleagris*). *Behaviour,* **1,** 237–66.

—— 1949: Das Verhalten gefangener Waldohreulen (*Asio otus otus*) und Waldkäuze (*Strix aluco aluco*) zur Beute. Ibid. **2,** 1–95.

RAND, A. L., 1941: Lorenz's objective method of interpreting bird behaviour. *Auk,* **58,** 289–91.

—— 1942: Nest sanitation and an alleged releaser. Ibid. **59,** 404–9.

—— 1943: Some irrelevant behavior in birds. Ibid. **60,** 168–71.

RHINE, J. B., and W. MCDOUGALL, 1933: Third report on a Lamarckian experiment. *Brit. J. Psychol.* **24,** 213–35.

RIDDLE, O., 1941: Endocrine aspects of the physiology of reproduction. *Ann. Rev. Physiol.* **3,** 573–616.

RINKEL, G. L., 1940: Waarnemingen over het gedrag van de kievit (*Vanellus vanellus* (L.)) gedurende de broedtijd. *Ardea,* **29,** 108–47.

ROWAN, W., 1932: Experiments in bird migration. III. The effects of artificial light, castration and certain extracts on the autumn movements of the American Crow (*Corvus brachyrhynchos*). *Proc. Nat. Acad. Sci.* **18,** 639–54.

—— 1938: Light and seasonal reproduction in animals. *Biol. Rev.* **13,** 374–402.

RUSSELL, E. S., 1934: *The Behaviour of Animals.* London.

—— 1943: Perceptual and sensory signs in instinctive behaviour. *Proc. Linn. Soc. London,* **154,** 195–216.

—— 1945: *The Directiveness of Organic Activities.* Cambridge.

SAND, A., 1937: The mechanism of the lateral sense organs of fishes. *Proc. Roy. Soc.* B, **123**, 472–95.

SANTSCHI, F., 1911: Le mécanisme d'orientation chez les fourmis. *Rev. Suisse Zool.* **19**, 117–34.

SCHARRER, E., 1935: Die Empfindlichkeit der freien Flossenstrahlen des Knurrhahns (*Trigla*) für chemische Reize. *Zs. vergl. Physiol.* **22**, 145–54.

SCHENKEL, R., 1947: Ausdrucks-Studien an Wölfen. *Behaviour*, **1**, 81–130.

SCHUYL, G., and L. and N. TINBERGEN, 1936: Ethologische Beobachtungen am Baumfalken, *Falco s. subbuteo* L. *J. Ornithol.* **84**, 387–434.

SEIFERT, R., 1930: Sinnesphysiologische Untersuchungen am Kiemenfuß, *Triops cancriformis*. *Zs. vergl. Physiol.* **11**, 386–464.

SEITZ, A., 1940: Die Paarbildung bei einigen Cichliden I. *Zs. Tierpsychol.* **4**, 40–84.

—— 1941: Die Paarbildung bei einigen Cichliden II. Ibid. **5**, 74–101.

SELOUS, E., 1906–7: Observations tending to throw light on the question of sexual selection in birds, including a day-to-day diary on the breeding habits of the Ruff (*Machetes pugnax*). *Zoologist* (4), **10**, 201–19, 285–94, 419–28; **11**, 60–5, 161–82, 367–80.

SHOEMAKER, H. H., 1939. Effect of testosterone propionate on behavior of the female canary. *Proc. Soc. Exp. Biol. Med.* **41**, 299–302.

SPERRY, R. W., 1941: The effect of crossing nerves to antagonistic muscles in the hind limb of the Rat. *J. comp. Neurol.* **75**, 1–19.

—— 1942: Transplantation of motor nerves and muscles in the forelimb of the Rat (cited after P. Weiss, 1941).

SPIETH, H., 1947: Sexual behavior and isolation in *Drosophila*. I. The mating behavior of species of the *willistoni* group. *Evolution*, **1**, 17–31.

STEINIGER, F., 1938: *Warnen und Tarnen im Tierreich.* Berlin.

SUMNER, F. B., 1934: Does 'protective coloration' protect? *Proc. Nat. Acad. Sci. Wash.* **20**, 559–64.

—— 1935*a*: Evidence for the protective value of changeable coloration in fishes. *Amer. Natural.* **69**, 245–66.

—— 1935*b*: Studies of protective color changes. III. Experiments with fishes both as predators and prey. *Proc. Nat. Acad. Sci. Wash.* **21**, 345–53.

SZYMANSKI, J. S., 1913: Ein Versuch, die für das Liebesspiel charakteristischen Körperstellungen und Bewegungen bei der Weinbergschnecke künstlich hervorzurufen. *Pflüger's Archiv*, **149**, 471–82.

THORPE, W. H., 1939: Further experiments on pre-imaginal conditioning in insects. *Proc. Roy. Soc.* **127**, 424–33.

—— 1943*a*: Types of learning in insects and other Arthropods. I. *Brit. J. Psychol.* (Gen. Section), **33**, 220–34.

—— 1943*b*: Types of learning in insects and other Arthropods. II. Ibid. **34**, 20–31.

—— 1944: Types of learning in insects and other Arthropods. III. Ibid. **34**, 66–76.

—— 1945: The evolutionary significance of habitat selection. *J. anim. Ecol.* **14**, 67–70.

TINBERGEN, L., 1935: Bij het nest van de torenvalk. *De Levende Natuur*, **40**, 9–17.

—— 1939: Zur Fortpflanzungsethologie von *Sepia officinalis* L. *Arch. Néerl. Zool.* **3**, 323–64.

TINBERGEN, N., 1932: Über die Orientierung des Bienenwolfes (*Philanthus triangulum* Fabr.). *Zs. vergl. Physiol.* **16**, 305–35.

TINBERGEN, N., 1935: Über die Orientierung des Bienenwolfes II. Die Bienenjagd. *Zs. vergl. Physiol.* **21,** 699–716.

—— 1936*a*: The function of sexual fighting in birds; and the problem of the origin of 'territory'. *Bird Banding,* **7,** 1–8.

—— 1936*b*: Zur Soziologie der Silbermöwe (*Larus a. argentatus* Pontopp). *Beitr. Fortpflanzungsbiol. Vögel,* **12,** 89–96.

—— 1936*c*: Eenvoudige proeven over de zintuigfuncties van larve en imago van de geelgerande watertor. *De Levende Natuur,* **41,** 225–36.

—— 1939*a*: The behavior of the Snow Bunting in spring. *Trans. Linn. Soc. N.Y.* **5.**

—— 1939*b*: On the analysis of social organization among vertebrates, with special reference to birds. *Amer. Midl. Natural.* **21,** 210–34.

—— 1940: Die Übersprungbewegung. *Zs. Tierpsychol.* **4,** 1–40.

—— 1942: An objectivistic study of the innate behaviour of animals. *Biblioth. biotheor.* **1,** 39–98.

—— 1947: Waarvoor gebruikt de geelgerande watertor zijn ogen? *De Levende Natuur,* **50,** 71–3.

—— 1948*a*: Social releasers and the experimental method required for their study. *Wilson Bull.* **60,** 6–52.

—— 1948*b*: Dierkundeles in het meeuwenduin. *De Levende Natuur,* **51,** 49–56.

—— 1949: De functie van de rode vlek op de snavel van de zilvermeeuw. *Bijdragen tot de Dierkunde,* **28,** 453–65.

—— and H. L. BOOY, 1937: Nieuwe feiten over de sociologie van de zilvermeeuwen. *De Levende Natuur,* **41,** 325–44.

—— and J. J. A. VAN IERSEL, 1947: 'Displacement reactions' in the Three-spined stickleback. *Behaviour,* **1,** 56–63.

—— and W. KRUYT, 1938: Über die Orientierung des Bienenwolfes (*Philanthus triangulum* Fabr.) III. Die Bevorzugung bestimmter Wegmarken. *Zs. vergl. Physiol.* **25,** 292–334.

—— and D. J. KUENEN, 1939: Über die auslösenden und die richtunggebenden Reizsituationen der Sperrbewegung von jungen Drosseln (*Turdus m. merula* L. und *T. e. ericetorum* Turton). *Zs. Tierpsychol.* **3,** 37–60.

—— B. J. D. MEEUSE, L. K. BOEREMA, and W. W. VAROSSIEAU, 1942: Die Balz des Samtfalters, *Eumenis* (= *Satyrus*) *semele* (L.). Ibid. **5,** 182–226.

—— and A. C. PERDECK, 1950: On the stimulus situation releasing the begging response in the newly hatched Herring Gull chick (*Larus a. argentatus* Pont.). *Behaviour,* **3,** 1–38.

TIRALA, L. G., 1923: Die Form als Reiz. *Zool. Jahrb. Allg. Zool. Physiol.* **39,** 395–442.

TOLMAN, E. C., 1932: *Purposive Behaviour in Animals and Men.* New York–London.

TOMITA, G., and H. TUGE, 1938: The development of behavior of the paradise fish. *J. Shanghai Sci. Inst.* (4), **3,** 269–78.

TSANG, YU-CHUANG, 1938: Hunger motivation in gastrectomized rats. *J. Comp. Psychol.* **26,** 1–18.

TUGE, H., 1938: Early behavior of embryos of an ovoviviparous fish, *Lebistes reticulosus*. *J. Shanghai Sci. Inst.* (4), **3,** 177–87.

UEXKÜLL, J. VON, 1921: *Umwelt und Innenwelt der Tiere.* Berlin.

VAN DER PLANK, F. L., 1934: The effect of infrared waves on tawny owls. *Proc. Zool. Soc. London,* 505–7.

Verwey, J., 1930*a*: Die Paarungsbiologie des Fischreihers. *Zool. Jahrb. Allg. Zool. Physiol.* **48,** 1–120.

—— 1930*b*: Einiges über die Biologie ostindischer Mangrovekrabben. Treubia 12, 169–261.

Vesey-Fitzgerald, B., 1947: The senses of bats. *Endeavour,* **6,** 36–41.

Wahl, O., 1933: Beitrag zur Frage der biologischen Bedeutung des Zeitgedächtnisses der Bienen. *Zs. vergl. Physiol.* **18,** 709–41.

Waldvogel, W., 1945: Gähnen als dienzephal ausgelöstes Reizsymptom. *Helv. Physiol. Acta,* **3,** 329–34.

Watson, J. B., 1908: The behavior of noddy and sooty terns. Ibid. **103,** 187–225.

—— and K. S. Lashley, 1915: Homing and related activities of birds. *Carn. Inst. Wash. Publ.* 211.

Weiss, P., 1941*a*: Self-differentiation of the basic patterns of coordination. *Comp. Psychol. Monogr.* **17,** 1–96.

—— 1941*b*: Autonomous versus reflexogenous activity of the central nervous system. *Proc. Amer. Philos. Soc.* **84,** 53–64.

—— 1941*c*: Does sensory control play a constructive role in the development of motor coordination? *Schweiz. Med. Wochenschr.* **71,** 591–5.

Wertheimer, M., 1922: Untersuchungen zur Lehre von der Gestalt I. *Psychol. Forschung,* **1,** 47–58.

—— 1923: Untersuchungen zur Lehre von der Gestalt II. Ibid. **4.**

Whitman, Ch. O., 1898: *Animal Behavior.* Woods Hole.

—— 1919: The behavior of pigeons. *Carnegie Inst. Wash. Publ.* **257,** 3, 1–161.

Wiesner, P. B., and N. M. Sheard, 1933: *Maternal Behavior in the Rat.* Edinburgh–London.

Wigglesworth, V. B., and J. D. Gillet, 1934: The function of the antennae in *Rhodnius prolixus* (Hemiptera) and the mechanism of orientation to the host. *J. Exp. Biol.* **11,** 120–39.

Wohlfahrt, Th.A., 1932: Anatomische Untersuchungen über das Labyrinth der Elritze (*Phoxinus laevis* L.). *Zs. vergl. Physiol.* **17,** 659–86.

Wolf, E., 1926: Das Heimkehrvermögen der Bienen. I. Ibid. **3,** 615–91.

—— 1928: Das Heimkehrvermögen der Bienen. II. Ibid. **6,** 221–54.

Wolfson, A., 1941: Light versus activity in the regulation of the sexual cycles of birds: the rôle of the hypothalamus. *Condor,* **43,** 125–36.

—— 1942: Regulation of spring migration in juncos. Ibid. **44,** 237–63.

—— 1945: The rôle of the pituitary, fat deposition, and body weight in bird migration. Ibid. **47,** 95–127.

Wunder, W., 1927: Sinnesphysiologische Untersuchungen über die Nahrungsaufnahme verschiedener Knochenfischarten. *Zs. vergl. Physiol.* **6,** 67–98.

—— 1933: Experimentelle Untersuchungen am Bitterling (*Rhodeus amarus*) während der Laichzeit. *Zool. Anz.* Suppl. Bd. **6,** 221–7.

Zitrin, A., 1942: Induction of male copulatory behavior in a hen following administration of male hormone. *Endocrinology,* **31,** 690.

AUTHOR INDEX

SUBJECT INDEX